Springer Tracts in Advanced Robotics

Volume 157

Frank Chongwoo Park, Mechanical Engineering Department, Seoul National University, Seoul, Korea (Republic of)

S. E. Salcudean, The University of British Columbia, Vancouver, BC, Canada

Roland Siegwart, LEE J205, ETH Zürich, Institute of Robotics & Autonomous Systems Lab, Zürich, Switzerland

Gaurav S. Sukhatme, Department of Computer Science, University of Southern California, Los Angeles, CA, USA

The Springer Tracts in Advanced Robotics (STAR) publish new developments and advances in the fields of robotics research, rapidly and informally but with a high quality. The intent is to cover all the technical contents, applications, and multidisciplinary aspects of robotics, embedded in the fields of Mechanical Engineering, Computer Science, Electrical Engineering, Mechatronics, Control, and Life Sciences, as well as the methodologies behind them. Within the scope of the series are monographs, lecture notes, selected contributions from specialized conferences and workshops, as well as selected PhD theses.

Special offer: For all clients with a print standing order we offer free access to the electronic volumes of the Series published in the current year.

Indexed by SCOPUS, DBLP, EI Compendex, zbMATH, SCImago.

All books published in the series are submitted for consideration in Web of Science.

Karen Bodie · Maximilian Brunner ·
Mike Allenspach

Omnidirectional Tilt-Rotor Flying Robots for Aerial Physical Interaction

Modelling, Control, Design and Experiments

Springer

Karen Bodie
Department of Mechanical and Process Engineering
ETH Zurich
Zürich, Switzerland

Maximilian Brunner
ETH Zurich
Zürich, Switzerland

Mike Allenspach
ETH Zurich
Zürich, Switzerland

ISSN 1610-7438 ISSN 1610-742X (electronic)
Springer Tracts in Advanced Robotics
ISBN 978-3-031-45496-7 ISBN 978-3-031-45497-4 (eBook)
https://doi.org/10.1007/978-3-031-45497-4

This Springer imprint is published by the registered company Springer Nature Switzerland AG
The registered company address is: Gewerbestrasse 11, 6330 Cham, Switzerland

Paper in this product is recyclable.

Foreword

At the dawn of the century's third decade, robotics is reaching an elevated level of maturity and continues to benefit from the advances and innovations in its enabling technologies. These all are contributing to an unprecedented effort to bringing robots to human environment in hospitals and homes, factories and schools; in the field for robots fighting fires, making goods and products, picking fruits and watering the farmland, saving time and lives. Robots today hold the promise for making a considerable impact in a wide range of real-world applications from industrial manufacturing to healthcare, transportation and exploration of the deep space and sea. Tomorrow, robots will become pervasive and touch upon many aspects of modern life.

The *Springer Tracts in Advanced Robotics (STAR)* is devoted to bringing to the research community the latest advances in the robotics field on the basis of their significance and quality. Through a wide and timely dissemination of critical research developments in robotics, our objective with this series is to promote more exchanges and collaborations among the researchers in the community and contribute to further advancements in this rapidly growing field.

The monograph by Karen Bodie, Maximilian Brunner and Mike Allenspach is based on the first author's doctoral thesis. The contents are organized in seven chapters and one appendix. The focus is on omnidirectional tilt-rotor flying robots for aerial physical interaction. Ranging from modelling to control, from design prototypes to experimental evaluation, the key concept is the ability to dynamically re-orient thrust vectors which opens the door to enhanced dynamic system capabilities and novel applications.

A nice blend of original theoretical and practical results in the field of aerial robotic manipulation, this volume is a great addition to the STAR series.

Naples, Italy
August 2023

Bruno Siciliano
STAR Editor

Preface

The study of tilt-rotor omnidirectional aerial robots and their application to aerial physical interaction tasks is a new and exciting domain of robotics. This book details an approach to developing such systems, and evaluating their efficacy as tools for a novel suite of robotics applications. Through modelling, control, prototype design and experimental evaluation, we seek to inspire a future of versatile and autonomous aerial manipulators. Recent developments in the field of fully actuated aerial robots have demonstrated the exceptional advantages of these systems for physical interaction over their underactuated counterparts. While the decoupling of translational and rotational dynamics serves as the basis for these advantages, the added ability to dynamically re-orient thrust vectors opens the door to a great expanse of possible morphologies, dynamic system capabilities and new applications. This book is based on the doctoral thesis of Karen Bodie at ETH Zurich's Autonomous Systems Lab completed in March 2022, under the supervision of Prof. Dr. Roland Siegwart.

The content of this book can be viewed as a set of tools to use in the development of aerial robots for physical interaction. The target audience is therefore students and researchers with a strong foundation in robotics concepts, who are aiming to develop new platforms according to the requirements of a certain aerial interaction task. Although some background concepts are presented in the appendix, it is expected that the reader has an understanding of kinematics and rigid body dynamics, as well as feedback control and optimization. The content of this book is multidisciplinary, and while sections can be referenced independently, a strong grasp on the system model is key to understanding the decisions taken in control and mechanical design sections.

We intend to guide the reader through the topic of omnidirectional aerial manipulators from the basic problem definition to the full physical realization of an aerial robot. We start by introducing the context and state of the art in an introductory chapter, then reflect on the problem definition of aerial physical interaction tasks, and the capabilities an aerial robot must possess to fulfil these tasks (Chaps. 1 and 2). In Chap. 3, we develop the system model, either as a single rigid body or a multi-body flying base, and include the option of a mounted parallel manipulator. The models developed here are then used in Chap. 4 for model-based control in 6 DOF, which considers interaction control as well as the control of an actuated manipulator arm. With theoretical components in place, Chap. 5 shifts to design of the prototype system, providing the reader with the tools to create an equivalent or novel platform. Then Chap. 6 presents a series of experiments that use the prototype to validate the control approaches from Chap. 4. Finally, Chap. 7 draws conclusions on the body of work presented here, and aims to inspire the reader with areas of future work.

Ranging from general modelling to control, design choices and complete system prototypes, the content of this work acts as a guide for envisioning and building innovative systems that will push the frontier of aerial manipulation.

Acknowledgements

Much of the work presented in this book has resulted from the close collaboration of many past and current members of the ASL and other labs at ETH Zurich. This form of intense collaboration is common (and often essential) in the field of robotics, and as such a number of acknowledgements are in order to recognize important contributors who are not listed as authors.

Many thanks to Roland Siegwart for the opportunity he has offered to so many students and employees at ASL. For his avid demonstration of confidence in our abilities as growing researchers, and his uncanny ability to foster a dynamic and collaborative research environment, we all profit greatly. Michael Pantic was a close collaborator throughout the chapters of this book, making important contributions in sensor and system integration, path planning, experimental testing and related analysis of results. His work enabled the extraction of probabilistic distance normals from a TOF sensor, and his local trajectory planning framework on mesh surfaces was critical to a number of experimental results. The concrete inspection content was completed in close collaboration with Patrick Pfändler from ETH's Institute for Building Materials (IfB) under the guidance of Prof. Dr. Ueli Angst. Born out of a shared ETH Grant between IfB and ASL, this ongoing concrete inspection project offered continual meaningful applications of our research, and the cross-disciplinary integration of innovative sensing equipment into our robotic technology. Important mechatronics contributions from Zachary Taylor were instrumental in constructing the first prototype platform. Markus Bühler and the machine shop staff were instrumental for their help in prototyping, and Michael Riner and Matthias Müller provided extensive support in system electronics. Elements of several master's student projects are also included in this book. The Master Thesis of Luca Rinsoz formed the basis of the morphology design optimization section. The initial prototype of the delta parallel manipulator and its design optimization is based on the Master's Thesis of Andrea Eusebi, co-supervised by Christian Lannegger. The innovative infinite-rotation tilting mechanism presented in Chap. 5 is based on the Semester Thesis of Philipp Göldlin, co-supervised by Mina Kamel.

Finally, we would like to recognize the innovative work of the 11 bachelor students who took part in the Voliro focus project, who conceived of the design and realized the first tilt-rotor hexacopter prototype at ASL. Their high-intensity effort over 9 months resulted in a valuable proof of concept for this type of omnidirectional system, and their successful demonstration led to the continued development of omnidirectional tilt-rotor platforms and further extensions to aerial manipulation at ASL.

Financial Support

The research included in this book received funding from ETH Research Grants, the National Center of Competence in Research (NCCR) on Digital Fabrication, NCCR Robotics and Armasuisse Science and Technology.

Zürich, Switzerland

Karen Bodie
Maximilian Brunner
Mike Allenspach

Contents

Abbreviations

2D	Two dimensional
3D	Three dimensional
6D	Six dimensional
ASIC	Axis-selective impedance control
ASL	Autonomous Systems Lab
BLDC	Brushless direct current
CAD	Computer-aided design
CAN	Controller area network
CoM	Centre of mass
CSE	Copper sulphate electrode
DoF	Degrees of freedom
EKF	Extended Kalman filter
ESC	Electronic speed controller
FPV	First person view
GNSS	Global navigation satellite system
IMU	Inertial measurement unit
LiPo	Linear quadratic regulator with integral action
LQRI	Linear quadratic regulator with integral action
MAV	Micro-aerial vehicle
MDT	Multi-directional thrust
mocap	Motion capture
MPC	Model predictive control
MSF	Multi-sensor fusion
NDT	Non-destructive testing
PD	Proportional-derivative
PI	Proportional-integral
PID	Proportional-integral-derivative
PNE	Projected Newton-Euler
PWM	Pulse width modulation
RMSE	Root mean squared error
RC	Radio control

ROS	Robot operating system
RTK	Real-time kinematic positioning
SQP	Sequential quadratic programming
SLAM	Simultaneous localization and mapping
TOF	Time-of-flight
TST	Total station theodolite
UDT	Uni-directional thrust
VI	Visual-inertial
VIO	Visual-inertial odometry
VTOL	Vertical take-off and landing

Chapter 1
Introduction

The presence of robotics and artificial intelligence in our daily lives has been increasing steadily and relentlessly since the rise of digital technology. Robotic systems now function as extensions of human beings, enhancing and parallelizing our ability to perform complex, difficult and mundane tasks. We design mobile robots, whether on the ground, underwater or in the air, to act as our sensors and tools, multiplying our effectiveness and extending our reach to remote locations on Earth and beyond. As we attempt to further explore, observe, build and maintain the world around us, we embed autonomy in these systems, giving them the ability to interpret sensory information and follow through with their assigned tasks without explicit human intervention.

In this chapter, we explore the context and history around aerial robotics by reviewing the state-of-the-art, and we motivate the case for aerial physical interaction with the current research gaps. We then describe our steps to close those gaps and finally provide an overall outline for the contents and organization of this book.

1.1 State-of-the-Art

Aerial robots have dominated the stage for robotic development in the public eye over the past decade. Inspired by the helicopter that was the subject of fervent development in the first half of the twentieth century [1], the robotics community embraced rotary-wing mechanical systems in the early 2000 s, nearly a century after their first proof of concept by the French bicycle maker Paul Cornu [2]. Intensive commercial efforts have since pushed rotary-wing aerial robots in a rapid transition from basic research to widespread use. A major part of this success story is due to the new opportunities that accompany a sensor-equipped robot with an unbounded 3D workspace in combination with agile flight capabilities. Rotary-wing MAVs are engaged for commercial applications such as security monitoring and visual inspection, equipped with

K. Bodie et al., *Omnidirectional Tilt-Rotor Flying Robots for Aerial Physical Interaction*, Springer Tracts in Advanced Robotics 157,
https://doi.org/10.1007/978-3-031-45497-4_1

Fig. 1.1 Technological development from **a** the first helicopter flight by Paul Cornu in 1907 and **b** a modern FPV commercial drone by Parrot Drone SAS

cameras or specialized sensors to monitor areas or detect damage in structures. Aerial transport and delivery is another active area of development, taking to the skies to avoid the chaos of ground-level traffic. These systems have also been embraced by the general public for photography and personal entertainment, as amateur pilots with FPV goggles fly 'drones' along the high mountain ridges of the Alps, extending their vision to engage in a thrilling experience without the risk of physical injury. Figure 1.1 illustrates the contrast of a century's technological development.

As the development of commercial rotary-wing MAVs matures, the research community around aerial robotics has moved its focus to address new challenges and opportunities that stem from the competencies of these systems. A major branch of aerial robotics research has shifted from aerial observation and navigation to aerial interaction and manipulation, with the ultimate goal of extending autonomous mobile manipulation to an unbounded workspace. To achieve this ambitious goal, we require a new breed of aerial robots which we refer to as *fully actuated*, and we need new methods to control them. We must further develop strategies for aerial interaction, and new methods for precise and dynamic end-effector tracking from a flying base.

In the following sections, we explore the motivation, state-of-the-art and research gaps of aerial manipulation divided into the following three categories, which are graphically shown in Fig. 1.2. *Fully actuated and omnidirectional MAVs* are developed to extend the capabilities of common underactuated flying systems, achieving important criteria for aerial manipulation. *Aerial Interaction* with fully actuated MAVs is a recent research topic that offers many new opportunities in the control of aerial robots. Finally, *macro-micro aerial manipulation* strategies develop new capabilities for improved speed and precision of the end-effector point for aerial interaction from a flying base.

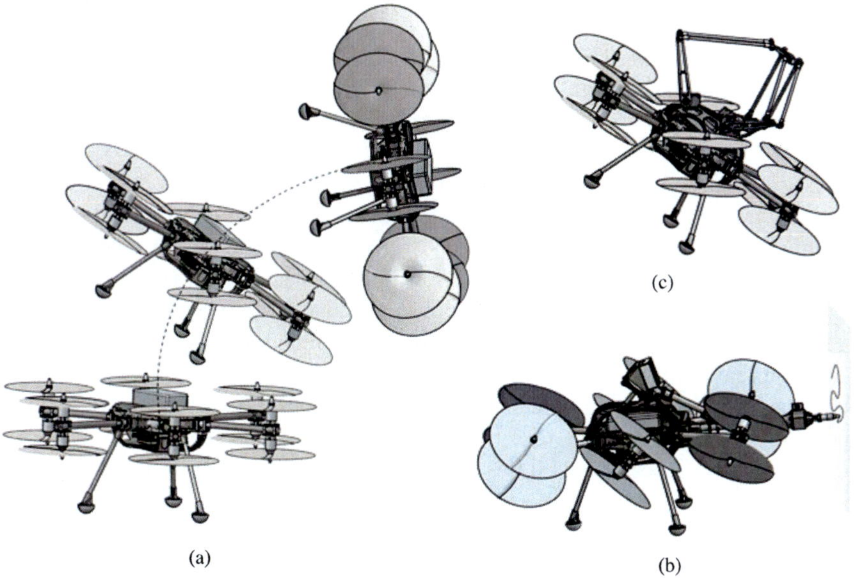

Fig. 1.2 Visual depictions of **a** an omnidirectional tilt-rotor platform and its applications for **b** aerial physical interaction and **c** macro-micro manipulation

1.1.1 Fully Actuated and Omnidirectional MAVs

Most commercial rotary-wing MAVs are underactuated UDT systems, providing only four controllable DOF by nature of their aligned propeller axes, regardless of the number of rotors. Underactuated systems such as these which emerged at the dawn of the twenty-first century [3] are highly dynamic in free flight with a simple system architecture where the only moving parts are the rotors themselves. This simplicity allows for highly representative dynamic modelling of the system, and aligned propeller axes further reduce unmodelled aerodynamic effects of airflow interaction between propellers. For aerial tasks that require exertion of arbitrary interaction forces and torques on the environment, or require the system to hold an arbitrary orientation outside of position and yaw, the limited controllable DOF of an underactuated system are no longer sufficient. Certain interaction tasks with underactuated systems can be performed [4, 5], but unexpected forces and torques applied to such a system during an interaction task can compromise tracking performance and lead to unstable behaviour.

A new class of *fully actuated* MAVs are now entering the aerial robotics curriculum [6], introducing systems with six independently controllable DOF. The resulting decoupled translational and rotational dynamics are critical to ensure stable interaction with the environment [7]. With the added criterion that the system's aerodynamic forces must be able to statically compensate the effects of gravity in any direction (i.e. maintain static hover in any base orientation), the result is an *omnidirectional* aerial robot, where a system can achieve uninhibited aerial movement and robust tracking of six DOF trajectories. This extension offers a unique advantage e.g. for aerial filming and 3D mapping in terms of uninhibited orientation of a sensor mounted on the body. Omnidirectional aerial platforms can further be used for configuration-based navigation in constrained environments, where rotating the vehicle body could allow for passage through a narrow gap [8]. Over the course of this work and in the few years before, substantial growth has occurred in the field of fully actuated and omnidirectional MAVs, from the proliferation of these systems in research labs to their application in realistic inspection scenarios. Fully actuated aerial systems present a compelling solution for future robotic applications, particularly in the emergent field of robotic inspection where physical interaction is required for tasks in remote or dangerous environments.

In the transition from simple underactuated MAVs to fully actuated and omnidirectional flying robots, the research community has been presented with a *new set of challenges*. With new capabilities come *increased system complexity*, demanding a more detailed system model, and control strategies that can tolerate new sources of model error. In addition to complexity, these systems suffer from *reduced efficiency* as new components add to the system mass, and variable thrust vectors create competing air flows that draw more power than UDT systems for an equivalent free-flight trajectory. A careful evaluation of the costs and benefits of these new systems is required to choose an appropriate platform design, and to justify its use for a given task. Drawing on prototypes of fully actuated flying robots from recently published research articles, we consider two dominant categories of platform actuation: fixed rotor and tilt-rotor systems. Images of some prominent state-of-the-art fully actuated systems are shown in Figs. 1.3 (fixed rotor) and 1.4 (tilt-rotor). Some additional systems step outside of these groups, exploring more unconventional morphologies, with examples depicted in Fig. 1.5.

The first category of *fixed rotor* systems is similar to the underactuated MAV in that the rotors are the only moving parts in the system. The difference from underactuated MAVs is that the propeller axes are no longer aligned, and their direction is chosen to optimize thrust vectoring capabilities for tasks requiring full six DOF

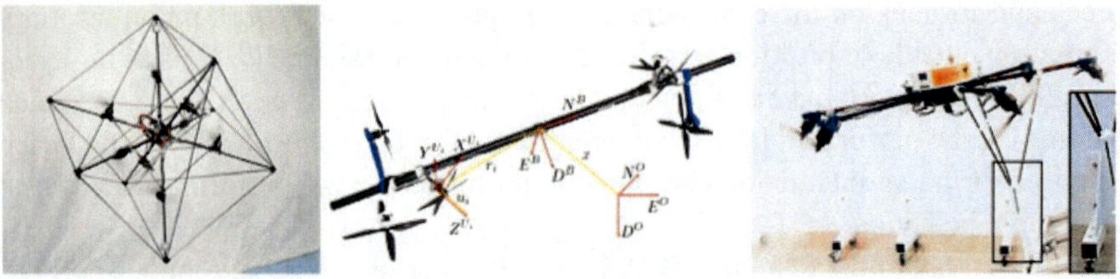

(a) (b) (c)

Fig. 1.3 State-of-the-art in fully actuated fixed rotor aerial robots. **a** An omnidirectional system from ETH Zurich [9], **b** ODAR, an omnidirectional system from SNU, Seoul [10], and **c** Tilt-Hex, a fully actuated system from LAAS-CNRS, Toulouse [11]

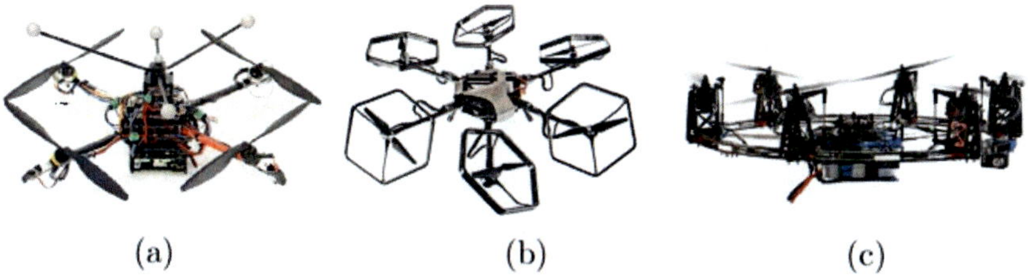

Fig. 1.4 State-of-the-art in fully actuated tilt-rotor aerial robots. **a** The 'holocopter', with four tilt-able rotor groups [24], **b** 'Voliro', an omnidirectional system with six tilt-able rotor groups from ETH Zurich [25], and **c** 'Fast-Hex', a system with tilt-able rotors coordinated by a single motor from the University of Twente [26]

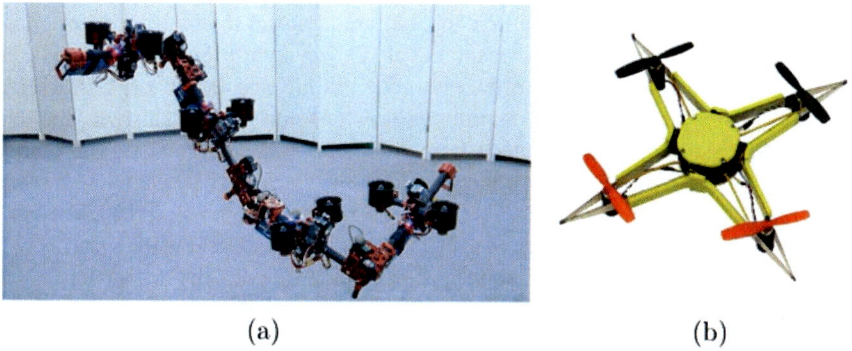

Fig. 1.5 Reconfigurable and flexible state-of-the-art aerial robots. **a** Chained modular dual-rotor links of the DRAGON aerial robot from University of Tokyo [36] and **b** an insect-inspired quad-rotor with integrated compliance from EPFL, Lausanne [37]

force and torque generation. These systems have the advantage that they change the direction and magnitude of thrust vector at a very high bandwidth, controlled by quickly adjusting propeller velocities, and are often designed such that the rotor forces can be inverted to increase omnidirectional force capabilities. By its design, however, a platform with fixedly tilted rotors that is able to generate a significant wrench on the environment has the disadvantage that it generates a proportionately significant amount of internal forces. These internal forces do not contribute to gravity compensation, and therefore directly detract from flight efficiency. Conversely, orienting the propellers to prefer efficient hover flight and a higher payload directly reduces the capability to generate lateral forces for interaction and disturbance rejection. Some general theoretical requirements for these systems are presented in [12]. An early example of a multi-rotor aerial system that stepped away from aligned propellers was a common UDT quad-rotor morphology with additional propellers on each arm [13, 14]. In this morphology, the secondary rotor axes were directed outwards from the centre of the body to generate lateral forces in order to decouple attitude and translation dynamics of the system. The research community continued this pursuit in the development of several hexacopter systems with non-collinear fixedly tilted rotors several years later, maintaining the traditional hexacopter structure and level flight [15–17]. This concept was later improved upon by Rajappa et al. [18], who optimized the rotor axes to minimize control effort and parasitic forces while ensuring the system's ability to perform a specified task. Based on this design optimization principle and preferring level flight, improved fixed rotor fully actuated platforms have been designed and realized, tolerating small amounts of roll and pitch while achieving omnidirectional wrench generation [19–21]. A new branch of fully-actuated fixed-rotor systems took off in 2016, with the presentation of two different experimental platforms that achieve full pose omnidirectionality [9, 22]. Propeller axes are optimized such that the system's flight properties remain similar in any hover orientation. This results in systems that generate high internal forces and therefore significantly reduced efficiency, but are agnostic to commanded orientations in space. In the work of Brescianini et al. [9], an eight-propeller configuration places each rotor group at the vertex of a cube within a protective cage structure, as a demonstration platform for omnidirectional trajectories [23]. This was the first experimental demonstration of unconstrained six DOF trajectory tracking for rotary-wing aerial robots, and was quickly followed in the same year by a similar system from Park et al. in Seoul [10]. In the work of Park et al. , both six [10] and eight-rotor [22] configurations have been experimentally demonstrated, with two arrays of rotors mounted to each end of a lightweight bar which supports the main electronics. Several key practical aspects of the physical system are developed, from smooth inversion of propeller spin direction to the first exploratory applications in physical interaction with an omnidirectional MAV.

The second category of *tilt-rotor* systems includes additional mechanisms to actively re-orient some or all of the propeller axes. Rotor groups can be tilted individually or collectively to re-orient the system's collective thrust vector. Thanks to this reconfigurability, tilt-rotor systems can achieve higher hover efficiency when lateral force exertion is not required, at the cost of additional inertia and mechanical complexity. An early example of a tilt-rotor system attempted to combine the attractive soaring properties of fixed-wing vehicles with the agility and hovering capabilities of a quad-rotor [27], a familiar concept that had been mechanically explored in the context of VTOL military aircraft. Conceptual development of the tilt-rotor quadcopter was introduced by Falconi et al. in 2012 [28], with the purpose of full actuation. The system's realization came soon after in the work of Ryll et al. [24, 29], with real flight experiments to validate the theory. The prototype system demonstrated full six DOF tracking capabilities, with limited roll and pitch due to actuation constraints. The design of a thrust-vectoring holonomic tilt-rotor tricopter was further explored in the work of Ramp et al. [30], realized physically in a number of other works [31, 32]. Adding a second degree of rotation at the main rotor groups in a compromise between omnidirectionality and force efficiency, Voliro AG has commercially pursued the concept of a thrust-vectoring tricopter [33]. The design relinquishes static rolling for highly efficient and unlimited pitching, maintaining a fully actuated platform with five DOF omnidirectionality. The first tilt-rotor system design capable of uninhibited omnidirectional flight was presented in the form of a tilt-rotor hexacopter by Kamel et al. [25]. With twelve actuators to control six degrees of freedom, this initial work validated the capabilities of the system in experimental flights, and presented a baseline PID controller and actuator allocation strategy for omnidirectional flight. Limitations of the system in terms of limited actuation capability in inefficient configurations, and open control challenges were presented as topics of future work. Another concept presented by Ryll et al. reduces the weight of six tilt motors, instead integrating a transmission system that couples all tilting rotor groups to a single motor [26, 34]. An extension to this work by Morbidi et al. includes a secondary tilt-axis which is orthogonal to the first [35]. Both of these coordinated-tilt solutions achieve versatility in force generation and efficiency in level flight, but the coupling of tilting rotor groups precludes omnidirectional flight and increases parasitic internal forces whenever lateral forces are required.

Morphology choice is practically unbounded in the context of aerial systems, and not all systems fall into the two general categories of fixed and tilt-rotor platforms. We can consider the previously mentioned tilt-rotor systems as a geometrically simple subset of reconfigurable aerial robots. More abstract concepts have been explored e.g. at the University of Tokyo, with the DRAGON [8, 36]: a fully actuated robot consisting of chained dual rotor links, where each connection point is articulated by two servomotors. The system offers some interesting advantages in its ability to pass through narrow gaps, and perform physical interaction tasks with each end of the chained system acting as a contact point. While most designs aim to reduce the effects of vibration and model error by designing components that approximate rigid bodies, some research begins to intentionally integrate compliance in flexible reconfigurable systems [37, 38]. Compliance can reduce damage to the system in case of a crash, or can be integrated to aid in interaction tasks.

Research gaps for fully actuated and omnidirectional aerial platforms are present in their design, modelling, control, and experimental evaluation. Since fully actuated systems are relatively new, the complexities of these new morphologies need to be carefully evaluated, particularly in the context of their control. Tilt-rotor systems have a high-dimensional actuation space which is vulnerable to *configuration-dependent singularities*. The identification, evaluation and robust handling of these conditions is an ongoing topic of research. Additional actuation also leads to a large number of actuation inputs to control a limited number of degrees of freedom, meaning that different combinations of actuator commands can result in the same motion or force generation of the aerial system. This *over-actuation* creates an opportunity to optimize control inputs based on secondary criteria such as efficiency or agility.

1.1.2 Aerial Physical Interaction

In parallel with the development of fully actuated aerial robots, the demand for aerial robotic workers for a wide range of physical interaction applications has been steadily gaining the attention of research communities, industry, and the general public [39, 40]. Aerial manipulation has had a strong practical appeal in research and industry for its promise to extend interactive and dexterous automated tasks to an unbounded workspace. Aerial robotic tasks can take on many forms, from the transportation of objects, to force application and modification of the surrounding environment.

The first major topic addressed in aerial manipulation was the *pick-and-place of objects* with quad-rotors, including the design of a gripping system and the modelling and control methods required to handle a variable payload [43, 44]. Similar work was occurring at the same time by Pounds et al. who investigated aerial transport and manipulation with autonomous helicopters [45, 46]. Helicopters at that time had the advantage of a much larger payload capacity and flight time, a gap that is closing as battery and actuator technology improves. Higher payload demands have been addressed through collaborative transportation of objects by a team of coordinated MAVs [47–50], and by the creation of much larger multi-rotor platforms [51, 52]. In the context of pick-and-place tasks, the major challenges for aerial manipulation are in precision for successfully grasping an object, and control of the platform with an added payload. Neither of these problems requires full actuation of the flying base, but may require an arbitrary pose of the gripper. Several systems have been developed to address the problem of underactuation by adding DOF in a manipulator arm with semi-static movement to reduce the dynamic complexity [39]. For underactuated aerial platforms which are able to control their position and yaw angle, a manipulator with at least two rotary DOF (e.g., a gimbal mechanism) is required to track a 6D end-effector trajectory.

A more recent topic in aerial interaction is sustained contact and force application on the environment, which we refer to as *aerial physical interaction*. In this case, the dynamics of the aerial system are coupled to the environment through physical contact. Techniques for ground-based [53, 54] and underwater [55] mobile manipulation matured in research before aerial robots first took flight. Now commercially used in manufacturing, these fields continue to develop in new research directions, particularly in autonomy for unstructured environments [56]. While some of these methods can be directly translated to aerial systems, a new set of challenges exists for interactive flying robots. Aerial physical interaction control has been demonstrated on several UDT aerial systems equipped with a manipulator [5, 42, 57], though the lateral force magnitudes and disturbance rejection capabilities of these systems during contact are limited due to the coupled rotational and translational dynamics of an underactuated flying base. Recent work by Lee et al. used a MPC approach with a UDT aerial system to open a door [58]. The task was performed successfully when the system was limited to low contact forces and slow lateral motion. While UDT systems may be ideal for pick-and-place tasks that require efficient transport of static payloads, sustained physical interaction in the presence of environmental disturbances is a task better suited for fully actuated aerial systems (Fig. 1.6).

A compelling and urgent example of the need for aerial physical interaction presents itself in the domain of *industrial contact inspection*, which has been a motivating interest in research and industry alike. The status of aging concrete infras-

(a) (b) (c)

Fig. 1.6 State-of-the-art in aerial physical interaction. **a** A fully actuated aerial manipulation platform equipped with a force sensor at the end-effector point from LAAS-CNRS [41], **b** an omnidirectional MAV with rigidly mounted end-effector from SNU in Seoul [22], and a UDT aerial system with an actuated end-effector from the University of Twente [42]

tructure is a growing concern due to the rising amount of required inspection, and a lack of capacity to meet the need by traditional means [59]. While embedded sensors in concrete structures are an ideal solution, this technology is only emerging now, and the next best approach to risk reduction for existing infrastructure is to perform early inspection. Early inspection promises a more efficient and intelligent approach to long-term maintenance, and a great cost savings when combined with automation. Technologies for NDT, such as potential mapping, permit detection of corrosion far earlier than visual assessment [60], but require sustained contact between the sensor and structure. As a result, these inspection methods require extensive human labour, road closure, and the use of large supporting inspection equipment, which is met in the worst case with insufficient maintenance budgets and shortages in qualified staff. The examples in Fig. 1.7 illustrate the costs and negative impact of inspection on both road users and inspection workers.

MAVs have been embraced as a solution for efficient visual inspection of infrastructure [61], and extending their capabilities to perform contact inspection is the next obvious step, which also comes with a new set of challenges. We now demand

(a) (b)

Fig. 1.7 Challenges in traditional bridge inspection: **a** a floating platform is used to support a truck inspecting the underside of a bridge, and **b** an inspection truck with a lifting platform requires lane closure and frequent repositioning, while workers lift inspection equipment repeatedly above their heads for the duration of the work the aerial robot to carry an additional (and often substantial) sensor payload and to exert precise forces on the environment in any direction, while at the same time rejecting other sources of disturbance. Fully actuated aerial robots are an ideal candidate for this realm of applications. In addition, new developments in inspection sensor technology have spawned several small and light-weight devices which make MAV-based inspection a feasible reality [62–65]. The combination of these emerging technologies leads to a promising new system architecture that is able to tackle the combined problems of interaction force control and disturbance rejection on an autonomous MAV.

The recent development of fully actuated aerial robots ensures that the flying base is capable of exerting an *arbitrary contact wrench* on the environment. If the system itself is omnidirectional, then a rigidly mounted end-effector extension is sufficient for 6D end-effector trajectory tracking and interaction with an arbitrarily oriented environment. This approach has been demonstrated for fixed rotor [22] and unconventional morphologies [36]. In the work of Park et al. [22], impressive demonstrations of aerial drawing, down-force application, and haptic peg-in-hole teleoperation were achieved. The work of Zhao et al. [36, 66] has demonstrated the ability of a multi-link reconfigurable flying robot to slide open the cover of a hatch and pass through the opening. A system that is fully actuated but *not* omnidirectional is able to track a 6D end-effector wrench, but requires additional degrees of freedom for full pose tracking. Several novel manipulators have been installed on fully actuated fixed rotor systems to enable full pose tracking at the end-effector [20, 21, 64, 67]. Staub et al. demonstrated robotic manipulation of an object connected by a hinge joint to the environment [21] as well as co-manipulation of an object in collaboration with a ground-based manipulator [67]. Push-and-slide contact inspection of a non-planar pipe surface was experimentally demonstrated by Tognon et al. [64] with the addition of a two DOF open-chain manipulator, while Ollero et al. demonstrated pipe inspection with a manipulator having one unconstrained rotational DOF and longitudinal compliance.

State-of-the-art *push-and-slide control approaches* for fully actuated systems vary in implementation: One approach uses cascaded PID control in free flight, switching to angular rate stabilizing control when in contact [42]. Another method implements a dislocated PD control law for an elastic jointed manipulator model, with integral action in all directions except along the tool's axis of contact [64]. A further approach tracks a pose trajectory on SE(3) in free flight, and switches to hybrid pose-wrench control when in contact [22]. Impedance control has been implemented on UDT aerial manipulators [68, 69], without omnidirectional force capabilities.

Handling of *external disturbances and contact forces* requires their observation, and has been successfully achieved on flying systems using momentum-based approaches [70, 71]. Tomic et al. further developed methods for model-based force estimation that discriminates contact forces from those caused by wind disturbances [72]. Several omnidirectional MAVs have performed contact inspection tasks on industrial structures [20, 64] without direct (closed-loop) force control. Implementations of *direct force feedback* on aerial systems have only emerged within the time-frame of this body of research (2017–2022), incorporating a force sensor mounted at the end-effector. In the work of Nava et al. , an optimal control approach uses force sensor readings to take advantage of contact forces in task execution [41], and in the work of Tzoumanikas et al. , a UDT system incorporates sensed contact forces in an MPC formulation for aerial writing [73]. Methods for direct interaction force control of fixed-base manipulators are well established [74, 75], typically switching controller modes when contact is detected. Switching controllers, however, are particularly unsuitable for flying systems due to the increased difficulty of contact estimation for a floating base in the presence of external disturbances. Recent improvements for force-controlled manipulators such as intelligent collision detection [76] and handling of contact loss during force control [77] are increasing commercial adoption. We can look to state-of-the art manipulator control techniques as inspiration for the control of newly capable fully actuated flying systems, keeping in mind the fundamental differences of a floating base system.

Aerial physical interaction with fully actuated vehicles presents new opportunities in the exploration of *passive controllers* with stability guarantees. A passivity control approach has been applied to a fully actuated aerial system for the first time by Rashad et al. using energy-tanks in a port-Hamiltonian formulation [78, 79]. Previously, passivity control techniques have been applied to UDT systems [4, 80], and extensively on fixed-base manipulator arms [81–83].

At the start of this work, a *research gap* in aerial manipulation platforms was present where omnidirectionality meets high force and torque capabilities, providing task versatility without the need for an actuated manipulator arm. Such a system maintains the simplified system dynamics of an omnidirectional flying base without compromising disturbance rejection capabilities. For control of fully actuated aerial manipulators for physical interaction, many methods have yet to be explored. We see the need for a controller that is used for all situations without reliance on explicit handling of contact transitions. A transition-tolerant controller will be essential for real-world interaction and situations where accidental contact should not compromise the task performance. Direct force feedback on an omnidirectional aerial manipulator had not been implemented at the beginning of the present work. While a few examples now exist in the literature, there is plenty of opportunity to implement previously developed force control methods on aerial systems for the first time, as well as entirely new control methods that address the specific qualities of aerial robots and their interaction with any object or environment.

1.1.3 Macro-Micro Aerial Manipulation

With the ultimate goal of performing complex manipulation tasks quickly and precisely in remote or dangerous environments, fully actuated aerial robots provide a strong basis on which to develop a new breed of interactive aerial workers. As these platforms are deployed into realistic environments, the effects of real-world disturbances on the system and implications on task performance must be taken into account. For omnidirectional robots with a fixed manipulator arm, disturbances to

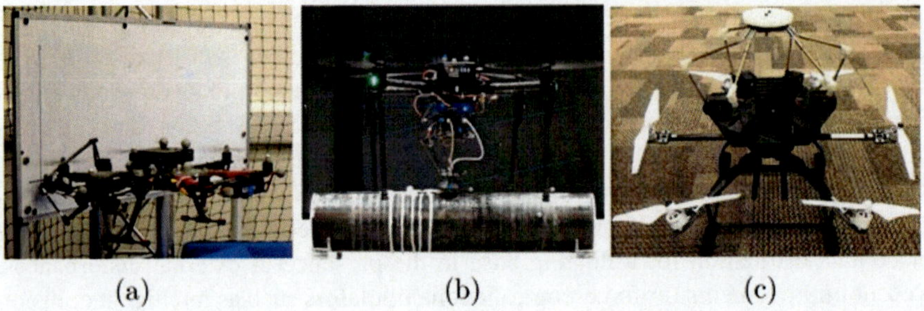

Fig. 1.8 State-of-the-art in macro-micro manipulation. **a** A UDT system with horizontally mounted parallel manipulator for aerial drawing from Imperial College London [73], **b** a UDT system with downward facing parallel manipulator for aerial repair from Imperial College London [85] and **c** a fully actuated system without omnidirectional flight equipped with a six DOF parallel manipulator from Purdue Polytechnic Institute [86]

the base are simultaneously present at the end-effector point, and the position error of the end-effector may be magnified by rotational error at the flying base. Fast end-effector operations may also demand dynamics higher than the floating base's actuation capabilities, leading to reduced performance or instability depending on the controller that is implemented. These drawbacks can be addressed by integrating local dynamic motion of the end-effector relative to the flying base within the platform's "infinite" workspace. Commonly referred to as the *macro-micro manipulator* model [84], it is sufficient to design a manipulator workspace and dynamics that can compensate for the natural fluctuations of the aerial base while providing a margin for task-specific dynamic motion.

To extend manipulation capabilities and overcome actuation limits of uni-, multi- and omnidirectional thrust vehicles, different *serial manipulator* designs, in single or double arm configuration, have been proposed [64, 65, 87]. To improve end-effector precision for point-contact tasks, UDT aerial vehicles have been equipped with *parallel manipulators* for operations below [85, 88, 89] or beside [73, 90–92] the flying robot. A fully actuated aerial system with a six DOF parallel manipulator mounted on top of the system is under development at Purdue Polytechnic Institute [86]. Some examples from the state-of-the-art are shown in Fig. 1.8.

For parallel aerial manipulators in literature up to this point, the manipulator has been considered as a pure kinematic system and controlled with a standard inverse kinematic method to compensate for positioning errors of the aerial platform. Dynamic coupling effects between the arm and the flying base are considered negligible, and are compensated by the aerial platform using the disturbance rejection capabilities of robust position controllers. This kinematic approach simplifies the control strategy, but performance degrades when rapid end-effector motion is required, and neglected manipulator dynamics manipulator become more significant. In extreme cases, these dynamics could lead to system instability.

For systems incorporating a multi-DOF arm on a flying base, different *control strategies* have been designed to handle the actuation redundancy. A baseline approach has been demonstrated through fully decoupled control of the base and end-effector [73, 85]. More sophisticated control techniques have been applied to fully actuated serial aerial manipulators including flatness-based [93] and full-body control [41].

In light of the current state-of-the art, a parallel manipulator mounted to a fully actuated or omnidirectional flying base has not been realized. Besides this, a *research gap* exists in incorporating parallel manipulator dynamics in control methods. In general, most aerial manipulation tasks have been performed slowly in a semi-static manner. Following the development path of fixed manipulators, incorporating dynamic movement in aerial manipulation tasks will lead to many exciting new opportunities in aerial physical interaction.

1.1.4 Contributions

Elements of this book advance the state-of-the-art, making steps to close the research gaps described above. We offer the following contributions in the topics of modelling, control, design, and experimental results.

Modelling

Modelling of MAVs is typically a simple affair considering a single rigid body with static transformations that relate propeller forces and torques to the centre of control of the platform. With tilt-rotor systems, we advance to a dynamic multi-body system. Whether or not these dynamics are significant enough to consider, omnidirectional flight with tilt-rotor systems brings with it a new set of *singularity conditions*, which we consider and evaluate in this work. Design of novel omnidirectional aerial systems can take many morphological forms, and optimization over the entire range of possibilities is a daunting task. In our work, we carefully select a set of parameters for omnidirectional tilt-rotor systems and present an optimization framework for *morphology optimization and evaluation*. The resulting platforms motivate the morphology choice for *a novel omnidirectional MAV* with versatile and highly dynamic performance capabilities.

Aerial manipulators have been considered in the past as dynamic multi-body extensions of a flying base. Parallel aerial manipulators have also been presented in literature, but their dynamics are neglected due to their low mass and relatively small movements. As we aim to push the end-effector speed and precision for a parallel aerial manipulator, we develop a model to incorporate the *dynamic effects of a floating base aerial manipulator*. The addition of a manipulator arm on the omnidirectional system is a non-trivial design task. In a compromise between inertial effects and the end-effector workspace, we pose a second optimization problem to select the geometric parameters for a delta manipulator.

In summary, we present the following contributions on the topic of modelling:

- Identification and evaluation of singularities of an omnidirectional tilt-rotor MAV [94].
- A tilt-rotor morphology optimization strategy to balance competing performance objectives [95].
- Formulation of the dynamic model of a floating base parallel manipulator, proposing relevant model simplifications validated by real experiments [96].

Control

With the new field of fully actuated and omnidirectional aerial robots come new opportunities in control for MAVs in 6 DOF. For tilt-rotor systems, we extend the state-of-the-art, taking *singular conditions* into account in the control framework to mitigate their negative effects. We further explore aerial interaction and disturbance rejection with the development of a *6 DOF axis-selective impedance controller*. Considering the transition between free flight and contact, we integrate *distance and direct force sensing* into the control framework to improve task performance in contact and more effective disturbance rejection in free flight, in a *hybrid force-impedance* control strategy. With the addition of a position-redundant manipulator on the omnidirectional flying base, we incorporate *parallel manipulator dynamics* and compensate these effects in the base controller. In summary, we present the following contributions on the topic of control:

- Actuator allocation strategies for avoiding the singularities of the tilt-rotor system [94, 95].
- A 6 DOF axis-selective impedance controller for a fully-actuated flying system [97].
- Hybrid force-impedance control in the form of direct force control combined with variable axis-selective impedance control for an omnidirectional aerial system [98].
- Control structure that integrates dynamic effects of an active manipulator with a feed forward dynamic compensation term [96].

Design

To demonstrate the theoretical capabilities of a flying robot concept in real flight experiments, a representative and reliable physical system is essential. We therefore detail the *design and construction* of several omnidirectional tilt-rotor systems, including the complete mechanical design and selection and integration of components for power, actuation, and computation. Fully on-board computation, power, and sensing are integrated, as well as a rigidly mounted arm with an integrated 6-axis force-torque sensor for aerial physical interaction. A collection of *system design considerations* presented over the contributing papers incorporate *novel elements for tilt-rotor aerial manipulation systems*. In summary, we present the following contributions on the topic of design, which support contributions in control and experimental evaluation:

- Design and prototype of a novel omnidirectional MAV with highly dynamic capabilities, while maintaining high efficiency in hover [94].
- The integration of a rigid manipulator arm, with direct force sensing at the end-effector, and a platform capable of on-board computation, power, and sensing [97, 98].
- Design considerations and prototype for the first implementation of a delta parallel manipulator on an omnidirectional flying robot [96].
- Hardware implementation considerations, as a resource for designing new omnidirectional systems [94–98].

Experimental Results

Experimental contributions are important for validating the proposed control techniques, including quantification of tracking error and qualification of the system behaviour. In order to present a *comparable and repeatable set of results*, we perform, evaluate, and present multiple trials of the same experimental flights. We further demonstrate that the system can operate outside of the controlled laboratory environment with on-board sensing. In summary, we present the following contributions on the topic of experimental results:

- Experimental performance validation of the prototype system and controllers for a tilt-rotor omnidirectional MAV in free flight [94, 95].
- Experimental validation of interaction control, and demonstration of the system as a viable platform for contact-based NDT of concrete infrastructure [97, 98].
- Experimental validation of the macro-micro manipulator strategy showing the performance improvement of a delta manipulator over a fixed arm for 6D end-effector trajectories [96].
- Experimental evaluation of manipulator dynamic compensation and resulting free-flight tracking performance [96].

1.2 Chapter Overview

A graphical overview of the book's organization is presented in Fig. 1.9, and details of the main content in each chapter are presented below.

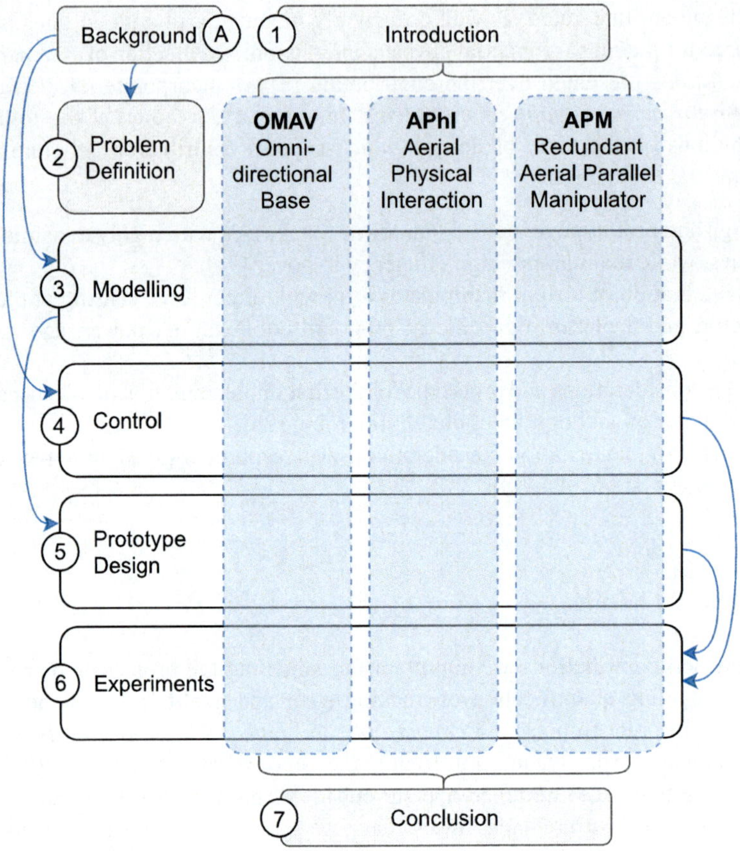

Fig. 1.9 Chapter overview

Problem Definition

Starting with a general problem definition, we define key requirements of aerial physical interaction tasks, the type of tasks that may emerge as this field grows, and the tasks that we aim to target in this body of work. Following this, we present key *properties of aerial systems*, how these capabilities align with requirements of a task, and how these properties can be represented and measured. We consider the capabilities of various morphologies of fully actuated flying robots and manipulators. Finally, we reflect on a number of related *research objectives*.

Modelling

Beginning with the simple model of a *rigid body flying base*, we develop a general system model that incorporates disturbances, interaction forces, and model errors. Next, we develop an *aerodynamic model* of the system, stating the assumptions of our system, and a joint model tilt-able rotor groups and actuated manipulator joints. The aerodynamic actuation wrench is developed and its properties are examined. An optimization problem is formulated to select parameters for the general *tilt-rotor model*, comparing the results of several cost functions to other fully actuated systems in the literature. Increasing complexity, we develop the model of a full tilt-rotor aerial system, and its extension to *aerial interaction* with a fixed manipulator. Finally we present the model of a *delta parallel manipulator* mounted to an omnidirectional flying base. A second optimization problem is formulated to select the parameters for a delta parallel manipulator, using a genetic algorithm. The detailed model of the parallel structure is considered as well as a set of simplifications to significantly reduce the model's complexity.

Control

We develop control approaches for an *omnidirectional flying base on SE(3)*, including the general control structure and generation of the 6 DOF control wrench. More specific to the tilt-rotor model, we present several methods for *actuator allocation* for the over-actuated flying system. With the integration of a rigidly mounted manipulator on the flying base, we develop methods for *interaction control*, first using an axis-selective *impedance control* method to enable compliant behaviour in the direction of contact while rejecting disturbances along the remaining axes. Using a force sensed at the end-effector, we then close a *direct force feedback* loop while in contact with the environment. Integration of a distance measurement is used to create a smooth transition between free flight and force-controlled interaction. The combination of both types of sensory feedback aims to improve the performance of contact tasks through *hybrid force-impedance* control. Finally we integrate a position-redundant delta parallel manipulator on the omnidirectional flying base and explore strategies for *macro-micro control* of redundant aerial manipulators. A baseline *decoupled controller* is developed, including handling of target points outside the limited manipulator workspace. With the goal of fast and precise tasks, we then integrate the manipulator dynamics in the controller as a *feed forward compensation* term.

Prototype Design

This chapter introduces the *detailed prototype design* and hardware implementation, including different options for basic structures, tilt-rotor actuation, rotor groups, manipulator structure, and mounted accessories. The *system electronics*, including power and communication, are described, and the general software framework is presented.

Experiments

We present an extensive suite of experiments which validate the proposed control approaches on our prototype system, and evaluate the performance and behaviour of these methods. The *experimental setup* is first described in detail, including different methods for state estimation, and the environments in which our flight tests are conducted. A series of flight experiments evaluate omnidirectional flight of the *tilt-rotor base*, focusing on its performance near singular configurations. We subsequently present *interaction control* experiments with a fixed manipulator, focusing on force application tasks, disturbance rejection, and contact transitions. A final section presents free flight experiments with a *position redundant parallel aerial manipulator* to evaluate the performance of a macro-micro manipulator strategy.

Conclusion

The final chapter of this book reflects on the accomplishments of the presented work and the significance of these contributions in pushing the state-of-the-art forward. Following this conclusion, we present an optimistic outlook on what the future offers in aerial physical interaction research.

References

1. Warner, E.P.: The problem of the helicopter (1920)
2. Crowe, J.: The problem of vertical flight: an examination of the characteristics of the helicopter in comparison with the autogiro and cyclogiro. Aircr. Eng. Aerosp. Technol. (1934)
3. Bouabdallah, S., Noth, A., Siegwart, R.: PID vs LQ control techniques applied to an indoor micro quadrotor. In: 2004 IEEE/RSJ International Conference on Intelligent Robots and Systems (IROS)(IEEE Cat. No. 04CH37566), vol. 3, pp. 2451–2456. IEEE (2004)
4. Yüksel, B., Secchi, C., Bülthoff, H.H., Franchi, A.: Reshaping the physical properties of a quadrotor through IDA-PBC and its application to aerial physical interaction. In: 2014 IEEE International Conference on Robotics and Automation (ICRA), pp. 6258–6265. IEEE (2014)
5. Alexis, K., Darivianakis, G., Burri, M., Siegwart, R.: Aerial robotic contact-based inspection: planning and control. Auton. Robot. **40**(4), 631–655 (2016)
6. Franchi, A.: Interaction control of platforms with multi-directional total thrust. In: Aerial Robotic Manipulation, pp. 175–189. Springer (2019)
7. Rashad, R., Goerres, J., Aarts, R., Engelen, J.B., Stramigioli, S.: Fully actuated multirotor UAVs: a literature review. IEEE Robot. & Autom. Mag. **27**(3), 97–107 (2020)
8. Zhao, M., Shi, F., Anzai, T., Chaudhary, K., Chen, X., Okada, K., Inaba, M.: Flight motion of passing through small opening by dragon: transformable multilinked aerial robot. In: 2018 IEEE/RSJ International Conference on Intelligent Robots and Systems (IROS), pp. 4735–4742. IEEE (2018b)
9. Brescianini, D., D'Andrea, R.: Design, modeling and control of an omni-directional aerial vehicle. In: 2016 IEEE International Conference on Robotics and Automation (ICRA), pp. 3261–3266. IEEE (2016)

10. Park, S., Her, J., Kim, J., Lee, D.: Design, modeling and control of omni-directional aerial robot. In: 2016 IEEE/RSJ International Conference on Intelligent Robots and Systems (IROS), pp. 1570–1575. IEEE (2016)
11. Franchi, A., Carli, R., Bicego, D., Ryll, M.: Full-pose tracking control for aerial robotic systems with laterally bounded input force. IEEE Trans. Robot. **34**(2), 534–541 (2018)
12. Tognon, M., Franchi, A.: Omnidirectional aerial vehicles with unidirectional thrusters: theory, optimal design, and control. IEEE Robot. Autom. Lett. **3**(3), 2277–2282 (2018)
13. Romero, H., Salazar, S., Sanchez, A., Lozano, R.: A new UAV configuration having eight rotors: dynamical model and real-time control. In: 2007 46th IEEE Conference on Decision and Control, pp. 6418–6423. IEEE (2007)
14. Salazar, S., Romero, H., Lozano, R., Castillo, P.: Modeling and real-time stabilization of an aircraft having eight rotors. In: Unmanned Aircraft Systems, pp. 455–470. Springer (2008)
15. Crowther, B., Lanzon, A., Maya-Gonzalez, M., Langkamp, D.: Kinematic analysis and control design for a nonplanar multirotor vehicle. J. Guid. Control. Dyn. **34**(4), 1157–1171 (2011)
16. Voyles, R., Jiang, G.: Hexrotor UAV platform enabling dextrous interaction with structures-preliminary work. In: 2012 IEEE International Symposium on Safety, Security, and Rescue Robotics (SSRR), pp. 1–7. IEEE (2012)
17. Jiang, G., Voyles, R.: Hexrotor UAV platform enabling dextrous interaction with structures-flight test. In: 2013 IEEE International Symposium on Safety, Security, and Rescue Robotics (SSRR), pp. 1–6. IEEE (2013)
18. Rajappa, S., Ryll, M., Bülthoff, H.H., Franchi, A.: Modeling, control and design optimization for a fully-actuated hexarotor aerial vehicle with tilted propellers. In: 2015 IEEE international conference on robotics and automation (ICRA), pp. 4006–4013. IEEE (2015)
19. Ryll, M., Bicego, D., Franchi, A.: A truly redundant aerial manipulator exploiting a multi-directional thrust base. IFAC-PapersOnLine **51**(22), 138–143 (2018)
20. Ollero, A., Heredia, G., Franchi, A., Antonelli, G., Kondak, K., Sanfeliu, A., Viguria, A., Ramiro Martinez-De Dios, J., Pierri, F., Cortés, J., Santamaria-Navarro, A., Trujillo, M.A., Balachandran, R., Andrade-Cetto, J., Rodriguez, A.: The aeroarms project: aerial robots with advanced manipulation capabilities for inspection and maintenance. IEEE Robot. Autom. Mag. (2018)
21. Staub, N., Bicego, D., Sablé, Q., Arellano, V., Mishra, S., Franchi, A.: Towards a flying assistant paradigm: the OTHex. In: 2018 IEEE International Conference on Robotics and Automation (ICRA), pp. 6997–7002. IEEE (2018)
22. Park, S., Lee, J., Ahn, J., Kim, M., Her, J., Yang, G.-H., Lee, D.: ODAR: aerial manipulation platform enabling omnidirectional wrench generation. IEEE/ASME Trans. Mechatron. **23**(4), 1907–1918 (2018)
23. Brescianini, D., D'Andrea, R.: Computationally efficient trajectory generation for fully actuated multirotor vehicles. IEEE Trans. Robot. **34**(3), 555–571 (2018)
24. Ryll, M., Bülthoff, H.H., Giordano, P.R.: A novel overactuated quadrotor unmanned aerial vehicle: modeling, control, and experimental validation. IEEE Trans. Control. Syst. Technol. **23**(2), 540–556 (2015)
25. Kamel, M., Verling, S., Elkhatib, O., Sprecher, C., Wulkop, P., Taylor, Z., Siegwart, R., Gilitschenski, I.: The voliro omniorientational hexacopter: an agile and maneuverable tiltable-rotor aerial vehicle. IEEE Robotics & Automation Magazine **25**(4), 34–44 (2018)
26. Ryll, M., Bicego, D., Giurato, M., Lovera, M., Franchi, A.: Fast-hex–a morphing hexarotor: design, mechanical implementation, control and experimental validation (2020). arXiv:2004.06612
27. Hancer, C., Oner, K.T., Sirimoglu, E., Cetinsoy, E., Unel, M.: Robust position control of a tilt-wing quadrotor. In: 49th IEEE Conference on Decision and Control (CDC), pp. 4908–4913. IEEE (2010)
28. Falconi, R., Melchiorri, C.: Dynamic model and control of an over-actuated quadrotor uav. IFAC Proc. Vol. **45**(22), 192–197 (2012)
29. Ryll, M., Bülthoff, H.H., Giordano, P.R.: Modeling and control of a quadrotor UAV with tilting propellers. In: 2012 IEEE International Conference on Robotics and Automation, pp. 4606–4613. IEEE (2012)

30. Ramp, M., Papadopoulos, E.: On modeling and control of a holonomic vectoring tricopter. In: 2015 IEEE/RSJ International Conference on Intelligent Robots and Systems (IROS), pp. 662–668. IEEE (2015)
31. Enya, E.L., Benjamin, T., Jeremy, G., Kian, N.M.T., Vengadashalapathy, L., Yoong, T., Srigrarom, S., et al.: Development of UGS tilt-rotor surveillance tricopter UAV. J. Appl. Sci. Eng. **18**(2), 177–186 (2015)
32. Lee, W.Y.J., Mehndiratta, M., Kayacan, E., Lee, W., Mehndiratta, M., Kayacan, E.: Fly without borders with additive manufacturing: a microscale tilt-rotor tricopter design. In: Proceedings of the 3rd international conference on progress in additive manufacturing (Pro-AM 2018), pp. 256–261 (2018)
33. V. AG. Voliro airborne robotics (2021). https://voliro.com/
34. Ryll, M., Bicego, D., Franchi, A.: Modeling and control of FAST-Hex: a fully-actuated by synchronized-tilting hexarotor. In: 2016 IEEE/RSJ International Conference on Intelligent Robots and Systems (IROS), pp. 1689–1694. IEEE (2016)
35. Morbidi, F., Bicego, D., Ryll, M., Franchi, A.: Energy-efficient trajectory generation for a hexarotor with dual-tilting propellers. In: 2018 IEEE/RSJ International Conference on Intelligent Robots and Systems (IROS), pp. 6226–6232. IEEE (2018)
36. Zhao, M., Anzai, T., Shi, F., Chen, X., Okada, K., Inaba, M.: Design, modeling, and control of an aerial robot dragon: a dual-rotor-embedded multilink robot with the ability of multi-degree-of-freedom aerial transformation. IEEE Robot. Autom. Lett. **3**(2), 1176–1183 (2018)
37. Mintchev, S., de Rivaz, S., Floreano, D.: Insect-inspired mechanical resilience for multicopters. IEEE Robot. Autom. Lett. **2**(3), 1248–1255 (2017)
38. Patnaik, K., Zhang, W.: Towards reconfigurable and flexible multirotors. Int. J. Intell. Robot. Appl. 1–16 (2021)
39. Ruggiero, F., Lippiello, V., Ollero, A.: Aerial manipulation: a literature review. IEEE Robot. Autom. Lett. **3**(3), 1957–1964 (2018)
40. Ollero, A., Tognon, M., Suarez, A., Lee, D., Franchi, A.: Past, present, and future of aerial robotic manipulators. IEEE Trans. Robot. (2021)
41. Nava, G., Sablé, Q., Tognon, M., Pucci, D., Franchi, A.: Direct force feedback control and online multi-task optimization for aerial manipulators. IEEE Robot. Autom. Lett. **5**(2), 331–338 (2019)
42. Wopereis, H., van de Ridder, L., Lankhorst, T.J., Klooster, L., Bukai, E., Wuthier, D., Nikolakopoulos, G., Stramigioli, S., Engelen, J.B., Fumagalli, M.: Multimodal aerial locomotion: an approach to active tool handling. IEEE Robot. & Autom. Mag. (2018)
43. Albers, A., Trautmann, S., Howard, T., Nguyen, T.A., Frietsch, M., Sauter, C.: Semi-autonomous flying robot for physical interaction with environment. In: 2010 IEEE Conference on Robotics, Automation and Mechatronics, pp. 441–446. IEEE (2010)
44. Mellinger, D., Lindsey, Q., Shomin, M., Kumar, V.: Design, modeling, estimation and control for aerial grasping and manipulation. In: 2011 IEEE/RSJ International Conference on Intelligent Robots and Systems, pp. 2668–2673. IEEE (2011)
45. Pounds, P.E., Dollar, A.M.: Stability of helicopters in compliant contact under PD-PID control. IEEE Trans. Robot. **30**(6), 1472–1486 (2014)
46. Pounds, P.E., Dollar, A.: Hovering stability of helicopters with elastic constraints. In: Dynamic Systems and Control Conference, vol. 44182, pp. 781–788 (2010)
47. Tagliabue, A., Kamel, M., Siegwart, R., Nieto, J.: Robust collaborative object transportation using multiple MAVs. Int. J. Robot. Res. **38**(9), 1020–1044 (2019)
48. Six, D., Briot, S., Chriette, A., Martinet, P.: Dynamic modelling and control of flying parallel robots. Control. Eng. Pract. **117**, 104953 (2021)
49. Sanalitro, D., Savino, H.J., Tognon, M., Cortés, J., Franchi, A.: Full-pose manipulation control of a cable-suspended load with multiple UAVs under uncertainties. IEEE Robot. Autom. Lett. **5**(2), 2185–2191 (2020)
50. Nguyen, H.-N., Park, S., Park, J., Lee, D.: A novel robotic platform for aerial manipulation using quadrotors as rotating thrust generators. IEEE Trans. Robot. **34**(2), 353–369 (2018)

51. G.E.I.T.C. Ltd. Ehang | UAM—passenger autonomous aerial vehicle (2021). https://www.ehang.com
52. Ong, W., Srigrarom, S., Hesse, H.: Design methodology for heavy-lift unmanned aerial vehicles with coaxial rotors. In: AIAA Scitech 2019 Forum, pp. 2095 (2019)
53. Yamamoto, Y., Yun, X.: Coordinating locomotion and manipulation of a mobile manipulator. In: [1992] Proceedings of the 31st IEEE Conference on Decision and Control, pp. 2643–2648. IEEE (1992)
54. Bellicoso, C.D., Krämer, K., Stäuble, M., Sako, D., Jenelten, F., Bjelonic, M., Hutter, M.: Alma-articulated locomotion and manipulation for a torque-controllable robot. In: 2019 International Conference on Robotics and Automation (ICRA), pp. 8477–8483. IEEE (2019)
55. Lapierre, L., Fraisse, P., Dauchez, P.: Position/force control of an underwater mobile manipulator. J. Robot. Syst. **20**(12), 707–722 (2003)
56. Blomqvist, K., Breyer, M., Cramariuc, A., Förster, J., Grinvald, M., Tschopp, F., Chung, J.J., Ott, L., Nieto, J., Siegwart, R.: Go fetch: mobile manipulation in unstructured environments (2020). arXiv:2004.00899
57. Scholten, J.L., Fumagalli, M., Stramigioli, S., Carloni, R.: Interaction control of an UAV endowed with a manipulator. In: 2013 IEEE International Conference on Robotics and Automation (ICRA), pp. 4910–4915. IEEE (2013)
58. Lee, D., Seo, H., Kim, D., Kim, H.J.: Aerial manipulation using model predictive control for opening a hinged door. In: 2020 IEEE International Conference on Robotics and Automation (ICRA), pp. 1237–1242. IEEE (2020)
59. ASCE: Infrastructure Report Card, p. 2017. ASCE Reston, VA (2017)
60. Angst, U.M.: Challenges and opportunities in corrosion of steel in concrete. Mater. Struct. **51**(1), (2018)
61. Chan, B., Guan, H., Jo, J., Blumenstein, M.: Towards UAV-based bridge inspection systems: a review and an application perspective. Struct. Monit. Maint. **2**(3), 283–300 (2015)
62. Pfändler, P., Bodie, K., Angst, U., Siegwart, R.: Flying corrosion inspection robot for corrosion monitoring of civil structures–first results. In: SMAR 2019-Fifth Conference on Smart Monitoring, Assessment and Rehabilitation of Civil Structures-Program, pp. We 4. SMAR (2019)
63. Isla, J., Cegla, F.: EMAT phased array: a feasibility study of surface crack detection. Ultrason. **78**, 1–9 (2017)
64. Tognon, M., Chávez, H.A.T., Gasparin, E., Sablé, Q., Bicego, D., Mallet, A., Lany, M., Santi, G., Revaz, B., Cortés, J., et al.: A truly redundant aerial manipulator system with application to push-and-slide inspection in industrial plants. IEEE Robot. Autom. Lett. **4**(2), 1846–1851 (2019)
65. Trujillo, M.Á., Martínez-de Dios, J.R., Martín, C., Viguria, A., Ollero, A.: Novel aerial manipulator for accurate and robust industrial NDT contact inspection: a new tool for the oil and gas inspection industry. Sens. **19**(6), 1305 (2019)
66. Zhao, M., Shi, F., Anzai, T., Okada, K., Inaba, M.: Online motion planning for deforming maneuvering and manipulation by multilinked aerial robot based on differential kinematics. IEEE Robot. Autom. Lett. **5**(2), 1602–1609 (2020)
67. Staub, N., Mohammadi, M., Bicego, D., Delamare, Q., Yang, H., Prattichizzo, D., Giordano, P.R., Lee, D., Franchi, A.: The tele-MAGMaS: an aerial-ground comanipulator system. IEEE Robot. Autom. Mag. (2017)
68. Lippiello, V., Ruggiero, F.: Exploiting redundancy in Cartesian impedance control of UAVs equipped with a robotic arm. In: 2012 IEEE/RSJ International Conference on Intelligent Robots and Systems (IROS), pp. 3768–3773. IEEE (2012)
69. Ruggiero, F., Cacace, J., Sadeghian, H., Lippiello, V.: Impedance control of VToL UAVs with a momentum-based external generalized forces estimator. In: 2014 IEEE International Conference on Robotics and Automation (ICRA), pp. 2093–2099. IEEE (2014)
70. Ruggiero, F., Cacace, J., Sadeghian, H., Lippiello, V.: Passivity-based control of VToL UAVs with a momentum-based estimator of external wrench and unmodeled dynamics. Robot. Auton. Syst. **72**, 139–151 (2015)

71. Tomić, T., Ott, C., Haddadin, S.: External wrench estimation, collision detection, and reflex reaction for flying robots. IEEE Trans. Robot. **33**(6), 1467–1482 (2017)
72. Tomić, T., Lutz, P., Schmid, K., Mathers, A., Haddadin, S.: Simultaneous contact and aerodynamic force estimation (s-CAFE) for aerial robots. Int. J. Robot. Res. **39**(6), 688–728 (2020)
73. Tzoumanikas, D., Graule, F., Yan, Q., Shah, D., Popovic, M., Leutenegger, S.: Aerial manipulation using hybrid force and position NMPC applied to aerial writing. In: Robotics: Science and Systems XVI (2020)
74. Khatib, O.: A unified approach for motion and force control of robot manipulators: the operational space formulation. IEEE J. Robot. Autom. **3**(1), 43–53 (1987)
75. Raibert, M.H., Craig, J.J., et al.: Hybrid position/force control of manipulators. J. Dyn. Syst. Meas. Control. **103**(2), 126–133 (1981)
76. Haddadin, S., De Luca, A., Albu-Schäffer, A.: Robot collisions: a survey on detection, isolation, and identification. IEEE Trans. Robot. **33**(6), 1292–1312 (2017)
77. Schindlbeck, C., Haddadin, S.: Unified passivity-based cartesian force/impedance control for rigid and flexible joint robots via task-energy tanks. In: 2015 IEEE International Conference on Robotics and Automation (ICRA), pp. 440–447. IEEE (2015)
78. Rashad, R., Engelen, J.B., Stramigioli, S.: Energy tank-based wrench/impedance control of a fully-actuated hexarotor: a geometric port-hamiltonian approach. In: 2019 International Conference on Robotics and Automation (ICRA), pp. 6418–6424. IEEE (2019b)
79. Rashad, R., Califano, F., Stramigioli, S.: Port-hamiltonian passivity-based control on se (3) of a fully actuated uav for aerial physical interaction near-hovering. IEEE Robot. Autom. Lett. **4**(4), 4378–4385 (2019)
80. Acosta, J.Á., Sanchez, M., Ollero, A.: Robust control of underactuated aerial manipulators via IDA-PBC. In: 53rd IEEE Conference on Decision and Control, pp. 673–678. IEEE (2014)
81. Ferraguti, F., Secchi, C., Fantuzzi, C.: A tank-based approach to impedance control with variable stiffness. In: 2013 IEEE International Conference on Robotics and Automation, pp. 4948–4953. IEEE (2013)
82. Dietrich, A., Wu, X., Bussmann, K., Ott, C., Albu-Schäffer, A., Stramigioli, S.: Passive hierarchical impedance control via energy tanks
83. Secchi, C., Ferraguti, F.: Energy optimization for a robust and flexible interaction control. In: 2019 International Conference on Robotics and Automation (ICRA), pp. 1919–1925. IEEE (2019)
84. Sharon, A., Hogan, N., Hardt, D.E.: The macro/micro manipulator: an improved architecture for robot control. Robot. Comput. Integr. Manuf. **10**(3), 209–222 (1993)
85. Chermprayong, P., Zhang, K., Xiao, F., Kovac, M.: An integrated delta manipulator for aerial repair: a new aerial robotic system. IEEE Robot. & Autom. Mag. **26**(1), 54–66 (2019)
86. Praveen, A., Yang, H., Jang, H., Voyles, R.M.: Aerial mobile manipulator system to enable dexterous manipulations with increased precision (2020). arXiv:2010.09618
87. Suarez, A., Heredia, G., Ollero, A.: Physical-virtual impedance control in ultralightweight and compliant dual-arm aerial manipulators. IEEE Robot. Autom. Lett. **3**(3), 2553–2560 (2018). https://doi.org/10.1109/LRA.2018.2809964
88. Haberfeld, G.B., Sun, D., Hovakimyan, N.: Stabilization and optimal trajectory generation for a compact aerial manipulation system with a delta-type parallel robot. In: 2018 International Conference on Unmanned Aircraft Systems (ICUAS), pp. 1091–1100. IEEE (2018)
89. Kamel, M., Alexis, K., Siegwart, R.: Design and modeling of dexterous aerial manipulator. In: 2016 IEEE/RSJ International Conference on Intelligent Robots and Systems (IROS), pp. 4870–4876. IEEE (2016)
90. Fumagalli, M., Stramigioli, S., Carloni, R.: Mechatronic design of a robotic manipulator for unmanned aerial vehicles. In: 2016 IEEE/RSJ International Conference on Intelligent Robots and Systems (IROS), pp. 4843–4848. IEEE (2016)
91. Danko, T.W., Chaney, K.P., Oh, P.Y.: A parallel manipulator for mobile manipulating UAVs. In: 2015 IEEE International Conference on Technologies for Practical Robot Applications (TePRA), pp. 1–6. IEEE (2015)

92. Fumagalli, M., Naldi, R., Macchelli, A., Forte, F., Keemink, A.Q., Stramigioli, S., Carloni, R., Marconi, L.: Developing an aerial manipulator prototype: physical interaction with the environment. IEEE Robot. Autom. Mag. **21**(3), 41–50 (2014). https://doi.org/10.1109/MRA.2013.2287454
93. Tognon, M., Yüksel, B., Buondonno, G., Franchi, A.: Dynamic decentralized control for protocentric aerial manipulators. In: 2017 IEEE International Conference on Robotics and Automation (ICRA), pp. 6375–6380. Singapore (2017)
94. Bodie, K., Taylor, Z., Kamel, M., Siegwart, R.: Towards efficient full pose omnidirectionality with overactuated mavs. In: International Symposium on Experimental Robotics, pp. 85–95. Springer (2018)
95. Allenspach, M., Bodie, K., Brunner, M., Rinsoz, L., Taylor, Z., Kamel, M., Siegwart, R., Nieto, J.: Design and optimal control of a tiltrotor micro-aerial vehicle for efficient omnidirectional flight. Int. J. Robot. Res. **39**(10–11), 1305–1325 (2020)
96. Bodie, K., Tognon, M., Siegwart, R.: Dynamic end effector tracking with an omnidirectional parallel aerial manipulator. IEEE Robot. Autom. Lett. **6**(4), 8165–8172 (2021)
97. Bodie, K., Brunner, M., Pantic, M., Walser, S., Pfändler, P., Angst, U., Siegwart, R., Nieto, J.: An omnidirectional aerial manipulation platform for contact-based inspection. In: Robotics: Science and Systems XV, vol. 15 (2019)
98. Bodie, K., Brunner, M., Pantic, M., Walser, S., Pfändler, P., Angst, U., Siegwart, R., Nieto, J.: Active interaction force control for contact-based inspection with a fully actuated aerial vehicle. IEEE Trans. Robot. (2020)

Chapter 2
Problem Definition

Starting with a general problem definition, we define key *requirements of aerial physical interaction tasks*, the type of tasks that may emerge as this field grows, and the tasks that we aim to target in this body of work. Following this, we present key *properties of aerial systems*, how these properties can be quantified in robotic systems, and how these capabilities align with the task requirements. We consider the capabilities of various morphologies of fully actuated flying robots and manipulators, and the addition of an actuated manipulator for improved end-effector performance. Finally, we reflect on some key *research objectives* to address the problem of aerial physical interaction, which will be developed further in Chaps. 3–5.

2.1 Notation

Throughout the document, we aim to keep consistent mathematical notation. From time to time, simplifications improve readability, and these will be made clear in the text, with common simplifications mentioned here.

For general mathematical structures, we adopt the following notation conventions. Vectors and matrices are represented respectively by lower- and uppercase letters in bold.

$$\boldsymbol{a} \in \mathbb{R}^n \qquad \boldsymbol{A} \in \mathbb{R}^{n \times m} \qquad \text{for } n, m \in \mathbb{N}^+ \tag{2.1}$$

where $\mathbb{N}^+$ represents natural numbers not including zero, i.e. positive integers.

Many variables will be expressed with respect to a certain coordinate frame. Frames are represented by capital letters, e.g. $\mathcal{F}_B$ which consists of an origin point O_B and the orthonormal basis formed by unit vectors $\boldsymbol{e}_x^B, \boldsymbol{e}_y^B, \boldsymbol{e}_z^B$. In their own frame, the unit vectors have the following meaning:

K. Bodie et al., *Omnidirectional Tilt-Rotor Flying Robots for Aerial Physical Interaction*, Springer Tracts in Advanced Robotics 157,
https://doi.org/10.1007/978-3-031-45497-4_2

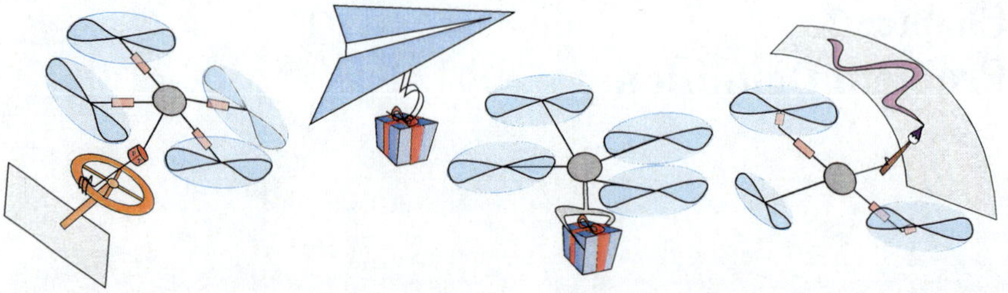

Fig. 2.1 Illustrations of select aerial physical interaction tasks, including modifying a constrained environment, object transport, pick-and-place, and push-and-slide aerial writing

$$\boldsymbol{e}_x = \begin{bmatrix} 1 \\ 0 \\ 0 \end{bmatrix} \quad \boldsymbol{e}_y = \begin{bmatrix} 0 \\ 1 \\ 0 \end{bmatrix} \quad \boldsymbol{e}_z = \begin{bmatrix} 0 \\ 0 \\ 1 \end{bmatrix} \tag{2.2}$$

The general inertial frame is also called the world frame, $\mathcal{F}_W$, and represents a frame that is not accelerating, and where the $\boldsymbol{e}_z^W$ axis is aligned with a gravitational body force $\boldsymbol{g} = g\boldsymbol{e}_z^W$, where $g = -9.81\text{ms}^{-2}$.

Common modifiers for vectors and matrices include $\cdot^*$ for a reference quantity , $\tilde{\cdot}$ for generalized forces and velocities, and $\bar{\cdot}$ for a modified vector or matrix that has a similar physical meaning. The time derivative of a vector $\boldsymbol{a}$ is generally written as $\dot{\boldsymbol{a}}$. For example, acceleration, velocity and position are related by time derivatives $\boldsymbol{a} = \frac{d}{dt}\boldsymbol{v} = \dot{\boldsymbol{v}}$ and $\boldsymbol{v} = \frac{d}{dt}\boldsymbol{p} = \dot{\boldsymbol{p}}$. The matrix $\mathbb{1}_n \in \mathbb{R}^{n \times n}$ represents the square identity matrix, and $\boldsymbol{0}_{m \times n} \in \mathbb{R}^{m \times n}$ is a matrix or vector populated by zeros. The subscript may be left out when the zero dimensions are implied by the context. We use other mathematical standard symbols, such as $|s|$ for the absolute value of a scalar s, and $||\boldsymbol{v}||$ for the Euclidean norm of a vector $\boldsymbol{v}$. A similar symbol is used to express a weighted vector norm, where

$$||\boldsymbol{u}||_{\boldsymbol{R}}^2 = \boldsymbol{u}^\top \boldsymbol{R} \boldsymbol{u}, \qquad \boldsymbol{u} \in \mathbb{R}^n, \ \boldsymbol{R} \in \mathbb{R}^{n \times n}. \tag{2.3}$$

A number of symbols used to express spatial pose and motion and dynamic quantities will be reused throughout this document. The most common of these are listed in Table 2.1.

Additional background material for understanding the kinematic and dynamic concepts used in this book is presented in Appendix. A section on kinematics reviews motion properties for rigid bodies, and their application in developing kinematic models of rigid multi-body systems. Formulation of the equations of motion for the dynamics of a rigid body system is reviewed, including the Lagrange and Newton-Euler formulations. These concepts will be used for modelling and control.

Table 2.1 Symbols and definitions for a tilt-rotor aerial vehicle

Symbol	Definition
$\mathcal{F}_W : \{O_W, \boldsymbol{e}_x^W, \boldsymbol{e}_y^W, \boldsymbol{e}_z^W\}$	Inertial frame: origin and primary axes
$\mathcal{F}_B : \{O_B, \boldsymbol{e}_x^B, \boldsymbol{e}_y^B, \boldsymbol{e}_z^B\}$	Robot body-fixed frame: origin and primary axes
$\mathcal{F}_{R_i} : \{O_{R_i}, \boldsymbol{e}_x^{R_i}, \boldsymbol{e}_y^{R_i}, \boldsymbol{e}_z^{R_i}\}$	ith rotor frame: origin and primary axes
${}^W\boldsymbol{p}_{AB}$	Position of point B w.r.t. A expressed in $\mathcal{F}_W$
${}^W\boldsymbol{v}_B$	Linear velocity of point B expressed in $\mathcal{F}_W$
$\boldsymbol{R}_{AB}$	Orientation of $\mathcal{F}_B$ expressed in $\mathcal{F}_A$
${}^B\boldsymbol{\omega}_{WB}$	Angular velocity of $\mathcal{F}_B$ w.r.t. $\mathcal{F}_W$ expressed in $\mathcal{F}_B$
m	Mass
$\mathcal{I}$	Inertia matrix
$\boldsymbol{f}$	Force vector
$\boldsymbol{\tau}$	Torque vector
$\boldsymbol{w}$	Wrench vector, $\boldsymbol{w} = [\boldsymbol{f}^\top \ \boldsymbol{\tau}^\top]^\top$
${}^W\boldsymbol{g} = [0\ 0\ g]^\top$	Gravity acceleration vector, $g = -9.81\text{ms}^{-2}$

2.2 Aerial Physical Interaction Tasks

We define *aerial interaction tasks* as objectives that are performed with an end-effector from an aerial base with measurable success criteria. Where *aerial* physical interaction sets itself apart from ground-based interaction is the unbounded workspace in which these tasks can be framed, and the ability of aerial vehicles to travel through free space with speed and agility. Considering the *reachable workspace* as the region of 3D points that the end-effector can reach and the *dextrous workspace* as the subset of these points that the end-effector can access in any orientation, aerial manipulators surpass fixed base manipulators in their ability to arbitrarily reposition their floating base. These advantages come at the cost of constant energy consumption to keep the vehicle and its payload in the air.

Task Descriptions

Physical interaction tasks vary in purpose, typically requiring a combination of precision, force application, and speed. *Pick-and-place* tasks require robustly grasping, transporting and releasing a certain object. Grasping and placement of the object may need to be precise, and may require force application, e.g. if the object must be mounted to a wall [1]. Transporting an object requires compensating its mass and inertia in the control scheme, and may involve obstacle detection and avoidance [2]. *Contact inspection tasks* can require stably applying a force to the environment in a precise location, and even a very specific force may be required. Continuous contact

Table 2.2 Aerial interaction tasks are compared based on the requirements they demand of an aerial platform, including precision of pose, force and torque tracking, magnitude of the required interaction wrench, speed of the operation, and omnidirectional interaction

	Precision			Wrench		Omni-
Task	Pose	Force	Torque	Magn.	Speed	direc.
Avoid Obstacles	◐	○	○	○	●	○
Grasp Object	●	◐	○	○	○	◐
Transport Object	○	○	○	○	●	○
Deposit Object	◐	◐	○	◐	○	◐
Point Inspection	◐	●	○	●	●	●
Rolling Inspection	◐	●	◐	●	◐	●
Aerial Drawing	●	◐	◐	◐	◐	◐
Modify Environment	◐	●	●	●	◐	●
Dextrous Manipulation	●	●	●	◐	●	●

inspection and drawing present some examples of *push-and-slide tasks*, where additional lateral forces are needed to overcome friction while maintaining a prescribed contact force into the surface. Precision of the end-effector point in the presence of surface friction and contact constraints may also be necessary. In some cases, tasks may require *modifying a constrained environment* such as turning a valve or opening a door. In such situations, a force and/or torque of a certain magnitude must be applied in a specific location and an arbitrary direction, and the dynamics of the environment become part of the task. *Dextrous manipulation* is the ultimate goal for aerial systems, meaning skilful competency with an end-effector to perform complex tasks. This type of task may demand precise movement of the end-effector in free space for complex interactions with the environment involving forces and torques. For all of the above-mentioned tasks, the performance value usually increases with the speed of operation. Higher speed permits more operations over time and also reduces the energy consumption required for a given operation, assuming that most energy is used to maintain hover of the aerial platform. Table 2.2 compares the relative high-level requirements for the tasks described above, and Fig. 2.1 illustrates some of these scenarios.

While grasping and transportation of objects have already been achieved to a very high-performance level, sustained contact of aerial systems with arbitrarily oriented surfaces is an open area of research. This book focuses on the cases of point inspection, rolling, and sliding along surfaces, and also makes preliminary steps in fast and precise end-effector motion for more complex tasks. The tasks of modifying an arbitrarily oriented environment, and performing dextrous manipulation tasks remain topics of future research.

Task Performance Metrics

Given that tracking precision and accuracy, arbitrary force and torque application, and speed are common elements of most physical interaction tasks, we take these as the important performance criteria for aerial manipulation systems to target. We can evaluate *tracking precision and accuracy* based on position and orientation error during a trajectory, both for the flying base and the end-effector point. Force error can be used to measure the performance of *force application*, considering both the transient and steady-state error values. Some tasks may require large forces and torques, so being able to exert an appropriate wrench magnitude for a task is also important. Minimizing time is best if it is the only factor, but we will also consider the compromise between *execution time* and the other performance factors. Measurement methods that will be used to quantify the three above-mentioned criteria are listed in Table 2.3.

While the above-mentioned performance criteria are quantifiable and measurable, the general behaviour of the system as it performs the task should also be taken into account, particularly in terms of repeatability and robustness to disturbances.

Table 2.3 Measurement methods and performance quantification are proposed for major aerial interaction task criteria. Error statistics include mean and standard deviation of the error as well as RMSE

Task criteria	Measurement	Quantification
Pose tracking	• state estimation	• pose error stats
	• ground truth pose	• pose error stats
Applied wrench	• wrench estimator	• force/torque error stats
	• force-torque sensor	• rise time, steady state offset
Execution time	• battery voltage	• total flight time
	• power consumption	• power per task activity

2.3 Properties of Aerial Manipulation Systems

Considering state-of-the-art aerial systems discussed in Sect. 1.1 and the wide variety of aerial vehicles that have been demonstrated, we simplify the morphology problem by considering only multirotor aerial systems which may contain rotary joints for thrust vectoring, where the control inputs, $\boldsymbol{u}$, are rotor speeds ($\boldsymbol{u}_r$) and joint angles ($\boldsymbol{u}_t$). Many other interesting morphology and actuation choices are possible, including the integration of variable-pitch propellers and airfoils, but lie outside the scope of this book.

The underlying factor that defines the performance limits of a rotary-wing system is the platform's physical actuation capabilities. Control approaches can be built to take advantage of these actuation features for improved flight performance, but will always be subject to their limitations. The precision of pose and motion tracking, as well as force application, are limited by the control bandwidth and capabilities of actuation forces and torques. The magnitude of interaction wrench that the system can exert on the environment, as well as the platform's ability to accelerate, are related to total maximum actuation forces and torques, in combination with the system's inertial properties.

For flying systems that require a large payload for on-board computation and sensing equipment, the system's large inertia reduces its ability to track highly dynamic motion. The concept of *macro-micro manipulation* offers a solution to improve task performance at the end-effector by adding a manipulator arm with faster dynamics. While this solution would be trivial for a fixed base, the manipulator choice is more subtle for a floating base system where reaction forces due to the arm dynamics can affect flight stability.

In this section, we first examine the *actuation capabilities of a flying base* by classifying them in terms of their degree of actuation, and defining metrics to measure and compare actuation across morphologies. We subsequently consider *properties of manipulators* which can be mounted to an aerial base, compare available morphologies, and examine how they can enhance the performance of the system.

2.4 Actuation Capabilities of a Flying Base

The actuation capabilities of a flying base are an important indication of their ability to perform an aerial interaction task. The *net actuation wrench*, ${}^{B}\boldsymbol{w}_{\text{act}}$, of the flying system is the sum of all actuation thrust forces and drag torques of the individual propeller units transformed in the floating base frame, computed as

$${}^{B}\boldsymbol{w}_{\text{act}} = \sum_{i=1}^{n_r} \boldsymbol{Ad}^{\top}_{\boldsymbol{T}^{-1}_{BR_i}} \boldsymbol{w}_{r_i} = \boldsymbol{A}_{\text{act}}\boldsymbol{u}_r, \qquad \boldsymbol{u}_r = \begin{bmatrix} f_{r_1} \\ \vdots \\ f_{r_{n_r}} \end{bmatrix} \tag{2.4}$$

where f_{r_i} is the individual thrust magnitude for the ith of n_r rotors. We refer to these thrust magnitudes as rotor inputs $\boldsymbol{u}_r$. For the remainder of this section we simplify notation, expressing the base frame actuation wrench, force and torque as $\boldsymbol{w}_{\text{act}}, \boldsymbol{f}_{\text{act}}, \boldsymbol{\tau}_{\text{act}}$.

This actuation wrench is used to control the motion of the system in order to track a trajectory in space as well a desired interaction wrench, and exists in a feasible wrench space $\mathbb{W}$ defined by rotor thrust limits

$$f_{r_{\min}} \leq f_{r_i} \leq f_{r_{\max}} \forall i \in \{1 \dots n_r\} \quad \rightarrow \quad \boldsymbol{u}_r \in \mathbb{U}_r \quad \rightarrow \quad \boldsymbol{w}_{\text{act}} \in \mathbb{W} \tag{2.5}$$

Control of the system further depends on the ability to change the actuation wrench. Taking the derivative of section eq. (2.4), we get

$$\dot{\boldsymbol{w}}_{\text{act}} = \frac{\partial \boldsymbol{w}_{\text{act}}}{\partial \boldsymbol{u}} \frac{\partial \boldsymbol{u}}{\partial t} = \dot{\boldsymbol{A}}_{\text{act}} \dot{\boldsymbol{u}}, \tag{2.6}$$

where the actuator input vector is $\boldsymbol{u} = \boldsymbol{u}_r$ if the system has fixed rotor positions, and $\boldsymbol{u} = [\boldsymbol{u}_r \ \boldsymbol{u}_t]$ for the rotor thrust and configuration angles, $\boldsymbol{u}_t$, of a tilt-rotor system. We refer to $\boldsymbol{A}_{\text{act}}$ as the *instantaneous actuator allocation matrix* and to $\dot{\boldsymbol{A}}_{\text{act}}$ as the *differential actuator allocation matrix*. For fixed rotor systems, these two matrices are equivalent, containing only constants. For tilt-rotor systems, these matrices are a function of the current tilt angle configuration $\boldsymbol{u}_t$, which are subject to angular position and velocity limits

$$\boldsymbol{u}_t \in \mathbb{U}_t, \qquad \dot{\boldsymbol{u}}_t \in \dot{\mathbb{U}}_t. \tag{2.7}$$

For the following definitions regarding the rank of $\dot{\boldsymbol{A}}_{\text{act}}$ for tilt-rotor systems, we are concerned with the existence of a tilt-angle configuration $\boldsymbol{u}_t \in \mathbb{U}_t$ that satisfies the proposed criteria.

System Classification by Actuation Capabilities

To define types of aerial systems based on their actuation capabilities, we use much of the terminology presented in [3], with some sight changes in definition. We first divide all rotary-wing aerial systems into *unidirectional thrust* (UDT) and *multidirectional thrust* (MDT) systems, where unidirectional systems have all rotors aligned as in the original quadrotor model, and multidirectional thrust contains all systems that are capable of some degree of thrust vectoring. *Fully actuated* systems are a subcategory of MDT, where the actuator allocation matrix achieves full rank. Multidirectional systems that do not fit this category include systems that can control their thrust vector along a plane, such as a tilt-rotor tricopter. Within the fully actuated category, systems can additionally be any of *omnidirectional* and *over-actuated*. Omnidirectional meaning that the system can hover in an arbitrary orientation without violating the limits of its actuators, and over-actuated indicating that an actuation

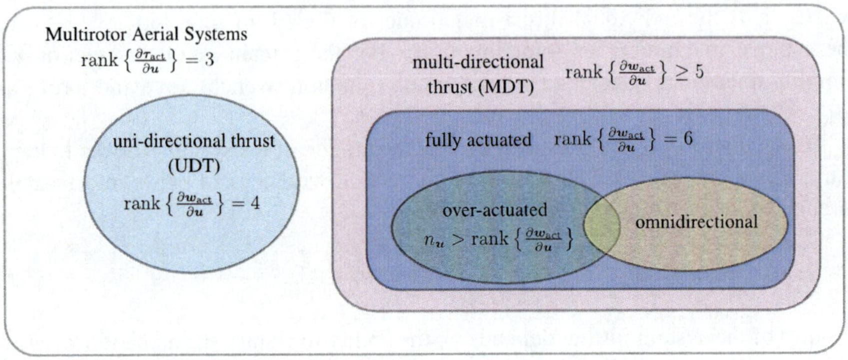

Fig. 2.2 Classification of aerial systems based on the allocation of the actuation inputs $\boldsymbol{u} \in \mathbb{R}^{n_u}$ to generate a net actuation wrench $\boldsymbol{w}_{act} = [\boldsymbol{f}_{act}^\top \ \boldsymbol{\tau}_{act}^\top]^\top$ in the flying base frame

null space exists due to an excess of control inputs compared to the DOFs of feasible actuation wrenches.

Mathematically, we complete these definitions as they relate to actuation inputs by evaluating the rank of the differential actuator allocation matrix, with a summary shown in Fig. 2.2. We consider rigid body systems with a time-invariant center of mass, which we choose as the origin for evaluating actuation wrenches. For further clarity in the following definitions, we refer to the force and torque components $\frac{\partial \boldsymbol{w}_{act}}{\partial \boldsymbol{u}} = [\frac{\partial \boldsymbol{f}_{act}}{\partial \boldsymbol{u}} \ \frac{\partial \boldsymbol{\tau}_{act}}{\partial \boldsymbol{u}}]^\top$. A basic assumption for all the multirotor systems considered here is that the torque component is full rank, meaning that the total moment can be generated and varied in any direction in $\mathbb{R}^3$:

$$\mathbf{rank}\left\{\frac{\partial \boldsymbol{\tau}_{act}}{\partial \boldsymbol{u}}\right\} = 3 \tag{2.8}$$

This assumption enables the basic attitude control of an aerial system. For the following definitions, we further require that there exists a feasible actuation wrench located at the COM with a non-zero force vector and zero-moment, such that in addition to re-orienting the base in space, the direction of linear motion can be controlled. The following requirements extend the above definitions as they relate to the rank of the actuator Jacobian matrix:

$$\underbrace{\mathbf{rank}\left\{\frac{\partial \boldsymbol{w}_{act}}{\partial \boldsymbol{u}}\right\} = 4}_{\text{Unidirectional Thrust}} \quad \underbrace{\mathbf{rank}\left\{\frac{\partial \boldsymbol{w}_{act}}{\partial \boldsymbol{u}}\right\} \geq 5}_{\text{Multidirectional Thrust}} \quad \underbrace{\mathbf{rank}\left\{\frac{\partial \boldsymbol{w}_{act}}{\partial \boldsymbol{u}}\right\} = 6}_{\text{Fully Actuated}} \quad \forall \boldsymbol{u} \in \mathbb{U} \tag{2.9}$$

For fully actuated systems, where the allocation is full rank, this implies that the net actuation force of the system can be varied independently of the net moment in all directions, decoupling the system's translational and rotational actuation and dynamics.

We refer to a subcategory of fully actuated systems as *omnidirectional*. For these systems, we first require that the net actuation thrust when represented at the COM can assume any direction in a spherical shell with zero moment, representing *force-omnidirectionality*. We additionally require the magnitude of this thrust to exceed the gravitational force in all directions, representing *body-omnidirectionality*:

$$||\boldsymbol{f}_{\text{act}}(\boldsymbol{u}^{\boldsymbol{d}*})|| > |g| \quad \forall\, \boldsymbol{d} \in \mathbb{R}^3 \text{ and } ||\boldsymbol{d}|| = 1 \tag{2.10}$$

where g is the gravitational constant, $\boldsymbol{d}$ is any direction vector in the unit sphere, and $\boldsymbol{u}^{\boldsymbol{d}*}$ are the actuator inputs that correspond to the maximum actuation force in direction $\boldsymbol{d}$.

In requiring the actuation force capability in any direction to compensate the effects of gravity, we push the definition of *omnidirectional* to mean a system that can sustain an omnidirectional pose in hover. At the purely analytical level, this requirement adds the need to estimate the system mass, which at a preliminary development stage may be unclear. As such, based on the actuator capabilities, we can instead identify the mass below which the system satisfies omnidirectionality.

We find this extension of body-omnidirectionality to be a more convenient definition for aerial platforms in two respects. First of all, fully actuated systems can become force-omnidirectional in the presence of gravity, where gravitational forces from the system's mass can be used as an actuation of sorts. For example, adding gravitational force to a system that can only exert forces upwards in a cone, the resulting envelope is force-omnidirectional in practice. Secondly, the extension to body-omnidirectionality adds a new meaning to this category of systems, allowing us to plan arbitrary waypoint poses for the flying base in six DOF.

A final classification is for *over-actuated* systems, where in addition to full actuation, the number of actuation inputs n_u exceeds the system's controllable DOFs. This means that different input combinations can be used to achieve the same net actuation wrench, indicating the presence of an actuation null space where secondary tasks may be performed.

$$\textbf{rank}\left\{\frac{\partial \boldsymbol{w}_{\text{act}}}{\partial \boldsymbol{u}}\right\} = 6 \quad \text{and} \quad n_u > \textbf{rank}\left\{\frac{\partial \boldsymbol{w}_{\text{act}}}{\partial \boldsymbol{u}}\right\} \tag{2.11}$$

Force and Torque Control Volumes

In order to compare the actuation capabilities of flying systems, we represent the wrench space $\mathbb{W}$ of a given morphology with *force and torque control volumes* of the system. We separate force and torque into separate 3D volumes that are centered in the floating base frame (nominally coincident with the center of mass) where the net actuation forces and torques are summed.

For a dense grid of vectors on the unit sphere, force volumes determine the maximum achievable pure force at the origin O_B along any axis, subject to $\boldsymbol{u}_r \in \mathbb{U}_r$.

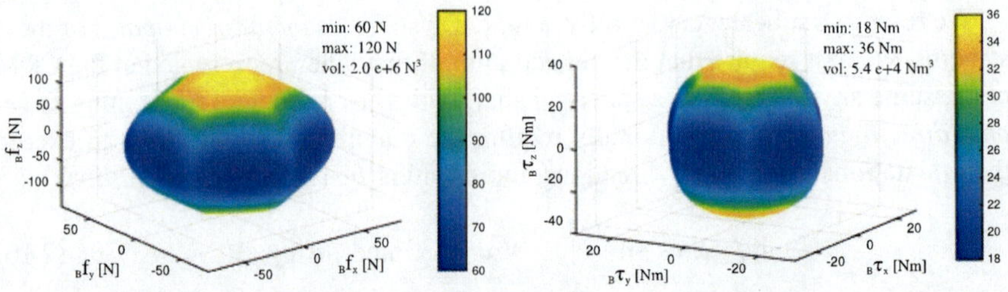

Fig. 2.3 Examples of force and torque volumes for a tilt-rotor hexacopter, where the surface colour corresponds with the force vector magnitude along the corresponding axis

Similarly, torque volumes evaluate the maximum achievable torque about a certain axis. They may also be expressed in the presence of a gravitational force, in which case the volume represents additional torques that can be applied while hovering in a specific orientation. An example of these volumes for a tilt-rotor hexacopter system is shown in Fig. 2.3.

Meaningful and comparable properties from force and torque volumes are the minimum and maximum magnitude, the average, and the total volume within the defined surface. For omnidirectional systems, the entire surface of the force envelope must lie outside of the gravitational sphere (i.e. a sphere of radius $m|g|$).

Efficiency and Parasitic Forces

A dominant struggle in the development of rotary-wing aerial systems is achieving a good compromise between performance and power consumption, or flight efficiency. The ability to quickly change a thrust vector may require the generation of competing forces, which consumes additional power for the same net control wrench at the flying base. These competing forces are referred to as *internal or parasitic forces* which can be used to measure the efficiency of the system compared to the hypothetical maximum where all rotors are aligned against gravity.

The *force efficiency* metric is defined as the ratio of the *net force magnitude* exerted by propellers in the base frame to the *sum of all the force magnitudes*, as originally formulated in [4]:

$$\eta_f(\boldsymbol{u}) = \frac{||\sum_{i=1}^{n_r} {}^B\boldsymbol{f}_{R_i}(f_{r_i}, \boldsymbol{q}_i)||}{\sum_{i=1}^{n_r} ||{}^B\boldsymbol{f}_{R_i}(f_{r_i}, \boldsymbol{q}_i)||} = \frac{||\sum_{i=1}^{n_r} {}^B\boldsymbol{f}_{R_i}(f_{r_i}, \boldsymbol{q}_i)||}{\sum_{i=1}^{n_r} f_{r_i}} \in [0, 1] \tag{2.12}$$

If all propeller forces are acting in the same direction, $\eta_f = 1$, indicating the most efficient configuration where no parasitic forces are present. Note that the system in this case is instantaneously equivalent to a UDT system due to the aligned propeller axes. With increased misalignment of propeller axes, internal forces grow, and the force efficiency decreases, to the extreme case where propeller forces cancel out exactly and $\eta_f = 0$.

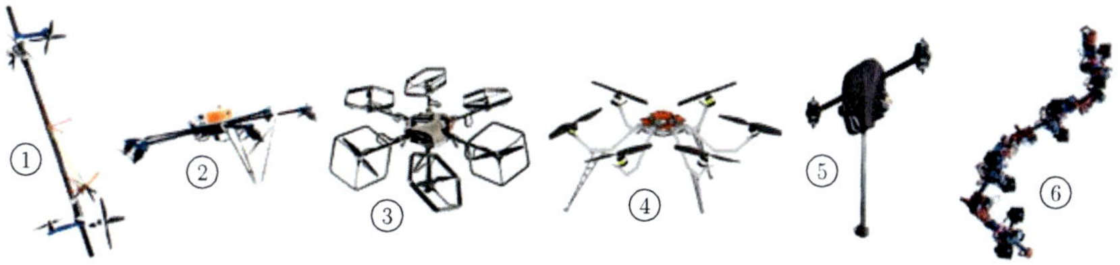

Fig. 2.4 Fixed and tilt-rotor fully actuated aerial platform morphologies inspired by state-of-the-art systems: (1) [5], (2) [6], (3) [7], (4) [8], (5) [9], (6) [10]

Capabilities by System Morphology

The literature survey in Sect. 1.1 touched on a wide variety of morphologies. Here we identify benefits and drawbacks of various implementations at a high level as they apply to aerial interaction tasks. We limit our evaluation to fully actuated flying systems consisting of rotor units that may be connected by rotary joints to one or more floating-base rigid bodies.

Inspired by state-of-the-art prototypes shown in Fig. 2.4, we describe and evaluate six morphologies which are defined as follows.

1. Fixedly-tilted motors are oriented to minimize performance variation during omnidirectional flight, inspired by [5]. Due to high internal forces, the system has low efficiency in flight, and marginal force exertion for interaction in directions that are not aligned with the gravity vector. As a system without tilt-able rotors, we consider this a simple system for which model error should be low, and reliability high.
2. Similar to the previous system, having fixed rotors that result in low system complexity, inspired by [6]. This system consists of rotors that are all placed on one plane, and forgoes omnidirectional flight to prefer flat hover, generating internal forces that are chosen to compromise efficiency and application of lateral forces.
3. Rotors groups lie on a plane, and are able to actively tilt individually about one axis, inspired by [7]. With some added complexity in tilt-actuation, we expect model accuracy to degrade. However, the additional tilt actuation allows for higher efficiency in body wrench commands since the actuators can be partially re-aligned in the desired direction, also allowing high magnitudes of force exertion.
4. Similar to the previous model, with an additional tilt mechanism for each motor such that each thrust vector can be chosen individually and exactly, inspired by simulation results from [8], though the physical system was not constructed. The complexity of implementing this system is high, and the mechanical construction and cable routing are very complex, with actuator limitations that would impede the theoretically perfect efficiency.
5. A fully actuated tricopter model inspired by [9]. Two thrust vectoring rotors enable highly efficient flight in five DOF, forsaking the ability to maintain roll in order

Table 2.4 Morphology comparison for fully actuated flying base

System	Complexity	Efficiency	Wrench magn.	Omni-force	Omni-pose
1	●	○	○	●	●
2	●	◐	◐	●	○
→ **3**	◐	◐	●	●	●
4	○	●	●	●	●
5	◐	●	●	●	◐
6	○	◐	◐	●	●

to achieve a simpler design. An exterior fixed motor enables large torques about the pitching axis.

6. A concept that achieves full pose omnidirectionality through reconfiguration of multiple rigid body core units, inspired by [10]. While interesting for its reconfigurability, its complexity and highly configuration-dependent efficiency and force exertion capabilities may detract from its ability to perform aerial physical interaction tasks.

To consider the suitability of different proposed morphologies for the tasks defined in Table 2.2, we translate the task criteria as they relate to actuation capabilities, and present high-level results in Table 2.4. Task precision in pose, force and torque relate to omnidirectionality via force and torque actuation envelopes. The wrench magnitude required by a task translates directly to that which the system can produce, beyond what is required to maintain hover. The speed of the system is related to its acceleration and therefore its actuation wrench. Flying systems in general can travel quickly across large distances, and an additional manipulator may be suitable for increased speed in a local workspace. Omnidirectionality of tasks relates in the minimal case to omnidirectional force application, and to omnidirectional-pose flight for the highest versatility by expanding the system's dexterous workspace. A new evaluation criteria of *complexity*, though not directly related to task criteria, relates to the physical system's number of moving parts and required actuators. With higher complexity, a system is more prone to component failure and model error, which ultimately affect the robot's ability to perform a task.

When designing a system to target the application of aerial workers, achieving a compromise between competing performance criteria is critical. Navigation efficiency and high payload capacity are often required, while general aerial physical interaction tasks demand high force and torque capabilities in all directions. The tilt-rotor omnidirectional MAV based on concept **3** offers a promising solution, having omnidirectional wrench exertion capabilities while maintaining the ability to revert to an efficient hover. While additional motors add complexity and weight, we see opportunity in the over-actuated system and its actuation null space. We select the general tilt-rotor system as a versatile base to address the problem of aerial manipulation, and will evaluate its theoretical and real-flight performance in the contents of this book.

2.5 Manipulators

For a fully actuated and omnidirectional flying base, a rigidly mounted static manipulator arm is sufficient for omnidirectional aerial manipulation. The actuation capabilities of the base are transferred to a point safely away from spinning propellers, to track omnidirectional motion and introduce force and torque on the environment.

The solution of a static arm retains the modelling simplicity of the base, but is also subject to the same actuation limits of the base transformed to the end-effector point. Therefore, the slow base dynamics of a heavy flying system influence the ability to perform dynamic tasks at the end-effector. In addition, error in the end-effector position is magnified with increasing base orientation error if a base reference is being tracked, and vice versa if tracking is framed at the end-effector point. In the best case, interaction properties at the end-effector are equivalent to or degraded from those of the flying base itself.

The addition of an actuated manipulator to an aerial system presents an opportunity to improve the system's performance characteristics at the end-effector point. While the force and torque generated by the flying base and its inertial properties are still the foundation of the system's capabilities, an actuated manipulator can further improve speed and precision within a limited workspace, according to the macro-micro principle: *"A macro/micro manipulator system, consisting of a large (macro) robot carrying a small (micro) high-performance robot, is proposed as a means of enhancing the functionality of a manipulator"* [11].

Enhanced Performance of Actuated Aerial Manipulators

When evaluating a combined system, we are interested in how task performance criteria can change with the addition of an active manipulator to an aerial system.

Defining $\boldsymbol{q}_m \in \mathbb{R}^{n_m}$ as a vector of n_m manipulator joint angles, we consider an end-effector Jacobian $\boldsymbol{J}_E$ for an actuated manipulator, which relates joint and end-effector velocities as

$$^{W}\boldsymbol{v}_B = \boldsymbol{J}_E(\boldsymbol{q}_m)\dot{\boldsymbol{q}}_m. \tag{2.13}$$

Fig. 2.5 Illustrations of serial and parallel manipulator concepts, with actuated and passive joints indicated respectively by shaded and empty circles, and approximate workspace volumes indicated by blue regions

For a manipulator with fast dynamics (dependent on geometry, morphology and actuator properties), these motion terms can quickly compensate for disturbances at the base, and trace end-effector trajectories that would be impossible for slower base dynamics to follow. Considering the desire for high interaction force control resolution, the manipulator may also be able to regulate interaction forces at a much higher bandwidth than the base controller. For this goal, it's important that the manipulator is able to sustain the desired interaction force at its joints, and to compensate at a high bandwidth with its own actuator torques. This relationship is also expressed with the manipulator Jacobian:

$$^{B}\boldsymbol{\tau}_{m} = \boldsymbol{J}_{E}^{\top}(\boldsymbol{q}_{m})\ ^{B}\boldsymbol{f}_{E} \tag{2.14}$$

Finally we consider the Jacobian condition number, $\kappa\left(\boldsymbol{J}_{E}\right)$, which represents the ratio of maximum and minimum eigenvalues. When the matrix is ill-conditioned, meaning there is a great difference in control authority on some dimension, we are near a singularity of the manipulator and will see reduced control performance.

Serial Verses Parallel Manipulators

The choice of manipulator morphology can greatly affect the resulting system performance, especially between open and closed chain structures. Although serial (open chain) manipulators have been the dominant choice in aerial manipulation because of their low complexity and large workspace, parallel (closed chain) manipulators offer a number of advantages that are particularly attractive when attached to an omnidirectional aerial base [12]. To compare these two major manipulator types for our task goals, we consider properties that affect end-effector position and force tracking performance (Fig. 2.5).

End effector error accumulates at each joint for serial link manipulators, where each successive error is added to the previous. The design of serial manipulators is therefore a game of high precision mechanics, which can be difficult to maintain in prototype aerial systems which may sustain damage in collisions or crashes. In parallel manipulators, errors at the joints that make up the closed chain structure are averaged at the end-effector point, improving the end-effector precision without introducing high costs or friction in the joint mechanics.

The inertia of moving parts is cascaded for serial manipulators, where each joint carries the weight of all successive joints towards the end-effector. This leads to significant dynamic effects on the system which must be accurately modelled if high accelerations are required at the end-effector. Parallel manipulators enable actuator placement at the platform base, leaving only the remaining links, passive joints and the end-effector as additional dynamic bodies to consider.

Effort distribution to the actuators is also cascaded for serial manipulators, driving the need for more powerful motors closer to the system base to sustain subsequent actuators as well as the arm's inertia. A force applied to the end-effector can generate a large moment arm that must be counteracted by any aligned actuator in the chain. In contrast, parallel manipulators distribute effort over the joints, combining the resulting forces over multiple chains. This also enables the selection of smaller actuators to generate the same end-effector forces.

The workspace of a serial manipulator is larger than that of a parallel manipulator when comparing total required link lengths. When selecting parameters based on required forces and precision at the end-effector, the comparison becomes more subtle, and largely a factor of the chosen components. The size of a workspace is less important than the resulting end-effector capabilities at any point, which can be evaluated by considering the condition number of $\boldsymbol{J}_E$ throughout the workspace.

The combination of properties listed above makes parallel manipulators particularly suitable for locally precise and dynamic motion of the end-effector within the platform's "infinite" workspace. Based on this conclusion, and the aforementioned simplicity of a rigidly mounted manipulator, we focus the contents of this book on static and parallel manipulators. In the work that follows, we use a rigidly mounted manipulator arm for physical interaction, or an actuated parallel manipulator to improve precision and speed at the end-effector as a first step towards dynamic interaction with the environment.

2.6 Objectives for Developing Robotics for Aerial Physical Interaction

The entire problem of aerial interaction spans several disciplines. In conceptual design and modelling, we seek to determine new and interesting capabilities of different system morphologies, and in control we seek high-performance, reliable and safe approaches for aerial physical interaction. A large body of work lies in the physical realization of aerial systems, including electronics and power integration, structure and mechanism design, and system maintenance while testing and crashing under various 'novel' controllers. While the implementation details of the hardware systems are minimally presented in research publications, these elements are critical in allowing further research in other domains such as control, perception, planning and multi-robot collaboration.

Through the work presented in this book, we aim to extend the capabilities of aerial manipulation systems for physical interaction in terms of system modelling, control, design and experimental evaluation, by the following methods:

Modelling and Conceptual Design

- Establish models of tilt-rotor systems and their corresponding assumptions.
- Identify performance criteria from models, and use optimization methods to find appropriate morphologies for omnidirectional aerial manipulation.

Control

- Motion control for an omnidirectional tilt-rotor flying base in SE(3).
- Interaction control for force application and contact transitions.
- Control of a redundant manipulator with dynamic compensation.

Prototype Design and Integration

- Design, construct and test fully actuated and omnidirectional tilt-rotor prototypes.
- Integrate physical extensions for force-controllable contact interaction integrating sensing and compliance.
- Design and construct a fast and light parallel aerial manipulator for precise end-effector tracking.

Experiments

- Demonstrate system performance through experimental flights.

References

1. Hamaza, S., Georgilas, I., Fernandez, M., Sanchez, P., Richardson, T., Heredia, G., Ollero, A.: Sensor installation and retrieval operations using an unmanned aerial manipulator. IEEE Robot. Autom. Lett. **4**(3), 2793–2800 (2019)
2. Oleynikova, H., Burri, M., Taylor, Z., Nieto, J., Siegwart, R., Galceran, E.: Continuous-time trajectory optimization for online uav replanning. In: 2016 IEEE/RSJ International Conference on Intelligent Robots and Systems (IROS), pp. 5332–5339. IEEE (2016)

3. Hamandi, M., Usai, F., Sablé, Q., Staub, N., Tognon, M., Franchi, A.: Design of multirotor aerial vehicles: a taxonomy based on input allocation. Int. J. Robot. Res. **40**(8–9), 1015–1044 (2021)
4. Ryll, M., Bicego, D., Franchi, A.: Modeling and control of FAST-Hex: a fully-actuated by synchronized-tilting hexarotor. In: 2016 IEEE/RSJ International Conference on Intelligent Robots and Systems (IROS), pp. 1689–1694. IEEE (2016)
5. Park, S., Her, J., Kim, J., Lee, D.: Design, modeling and control of omni-directional aerial robot. In: 2016 IEEE/RSJ International Conference on Intelligent Robots and Systems (IROS), pp. 1570–1575. IEEE (2016)
6. Ryll, M., Bicego, D., Franchi, A.: A truly redundant aerial manipulator exploiting a multi-directional thrust base. IFAC-PapersOnLine **51**(22), 138–143 (2018)
7. Kamel, M., Alexis, K., Siegwart, R.: Design and modeling of dexterous aerial manipulator. In: 2016 IEEE/RSJ International Conference on Intelligent Robots and Systems (IROS), pp. 4870–4876. IEEE (2016)
8. Rajappa, S., Ryll, M., Bülthoff, H.H., Franchi, A.: Modeling, control and design optimization for a fully-actuated hexarotor aerial vehicle with tilted propellers. In: 2015 IEEE International Conference on Robotics and Automation (ICRA), pp. 4006–4013. IEEE (2015)
9. AG, V.: Voliro airborne robotics (2021).https://voliro.com/
10. Zhao, M., Anzai, T., Shi, F., Chen, X., Okada, K., Inaba, M.: Design, modeling, and control of an aerial robot dragon: a dual-rotor-embedded multilink robot with the ability of multi-degree-of-freedom aerial transformation. IEEE Robot. Autom. Lett. **3**(2), 1176–1183 (2018)
11. Sharon, A., Hogan, N., Hardt, D.E.: The macro/micro manipulator: an improved architecture for robot control. Robot. Comput.-Integrat. Manufact. **10**(3), 09–222 (1993)
12. Tsai, L.-W.: Robot Analysis: The Mechanics of Serial and Parallel Manipulators. Wiley (1999)

Chapter 3
Modelling

To model a fully actuated tilt-rotor aerial manipulator, we start by considering a generic system of rigid bodies, with n DOF, represented by a set of generalized coordinates, $\boldsymbol{q}$, which uniquely and fully describe the system's physical configuration. The motion of the system is described as the evolution of the physical configuration over time, and is represented by generalized velocities, $\widetilde{\boldsymbol{u}}$, related to the time derivative of the generalized coordinates, $\dot{\boldsymbol{q}}$, as

$$\widetilde{\boldsymbol{u}} = \boldsymbol{J}_q(\boldsymbol{q})\dot{\boldsymbol{q}} \in \mathbb{R}^n \tag{3.1}$$

In the case that $\boldsymbol{J}_q = \mathbb{1}_n$, the generalized velocities are exactly the time derivatives of the coordinates $\boldsymbol{q}$. For a floating base, this formulation is complicated by our choice to represent the system's orientation on SO(3), which will be addressed in Sect. 3.2.

According to the Lagrangian formulation for dynamic systems (see Sect. A.2.1), the system's motion is defined by its inertial properties, the configuration, and the sum of external generalized forces acting on the system. We express the equations of motion in terms of generalized velocities, $\widetilde{\boldsymbol{u}}$, and their corresponding generalized forces $\widetilde{\boldsymbol{\tau}} \in \mathbb{R}^n$.

$$\sum \widetilde{\boldsymbol{\tau}} = \boldsymbol{\mathcal{M}}(\boldsymbol{q})\dot{\widetilde{\boldsymbol{u}}} + \boldsymbol{\mathcal{C}}(\boldsymbol{q}, \widetilde{\boldsymbol{u}})\widetilde{\boldsymbol{u}} + \boldsymbol{g}(\boldsymbol{q}) \tag{3.2}$$

where $\boldsymbol{\mathcal{M}} \in \mathbb{R}^{n\times n}$ is the configuration-dependent generalized inertia matrix, $\boldsymbol{\mathcal{C}} \in \mathbb{R}^{n\times n}$ is the matrix of nonlinear terms including centrifugal and Coriolis effects, and $\boldsymbol{g} \in \mathbb{R}^n$ is the vector of generalized gravitational forces.

For our floating base aerial robot, we model the generalized forces acting on the system in three separate categories as **actuation forces**, $\widetilde{\boldsymbol{\tau}}_{\text{act}}$, resulting from actuator inputs, $\boldsymbol{u}$, **external contact forces**, $\widetilde{\boldsymbol{\tau}}_{\text{ext}}$, originating from a contact wrench at the end-effector, and disturbance forces, $\widetilde{\boldsymbol{\tau}}_{\Delta}$, resulting from air currents and all other accumulated model error.

$$\sum \widetilde{\boldsymbol{\tau}} = \widetilde{\boldsymbol{\tau}}_{\text{act}}(\boldsymbol{u}) + \widetilde{\boldsymbol{\tau}}_{\text{ext}} + \widetilde{\boldsymbol{\tau}}_{\Delta} \tag{3.3}$$

K. Bodie et al., *Omnidirectional Tilt-Rotor Flying Robots for Aerial Physical Interaction*, Springer Tracts in Advanced Robotics 157,
https://doi.org/10.1007/978-3-031-45497-4_3

The rest of this chapter goes deeper into the details of this model, and different ways that we can represent the system, to use these models in subsequent control schemes.

Starting with Sect. 3.1, we develop a kinematic description of a tilt-rotor MAV. Based on an individual propeller model, we then formulate the net *aerodynamic actuation* wrench and examine its properties. Section 3.2 introduces the *tilt-rotor omnidirectional flying base* as a dynamic system, at different degrees of simplicity. An optimization problem then evaluates the general tilt-rotor system to select an appropriate morphology for omnidirectional aerial manipulation. In Sect. 3.3 we briefly discuss the aerial *interaction model* for a contact point at the end-effector including assumptions about the environment. Finally, in Sect. 3.4 we develop the model of a *delta parallel manipulator* on a floating base (macro-micro model), and formulate a genetic optimization problem to select the system's geometric parameters. We then describe the manipulator dynamics and propose model simplifications.

Figure 3.1 shows a brief overview of the different model representations at varying levels of simplicity, subject to actuation and contact forces described in Eq. (3.3), with coordinates for a floating base, $\boldsymbol{q}_b$, tilting rotor groups, $\boldsymbol{q}_t$, and actuated manipulator joints, $\boldsymbol{q}_m$. In this book, only the single-body base model is considered for subsequent control methods.

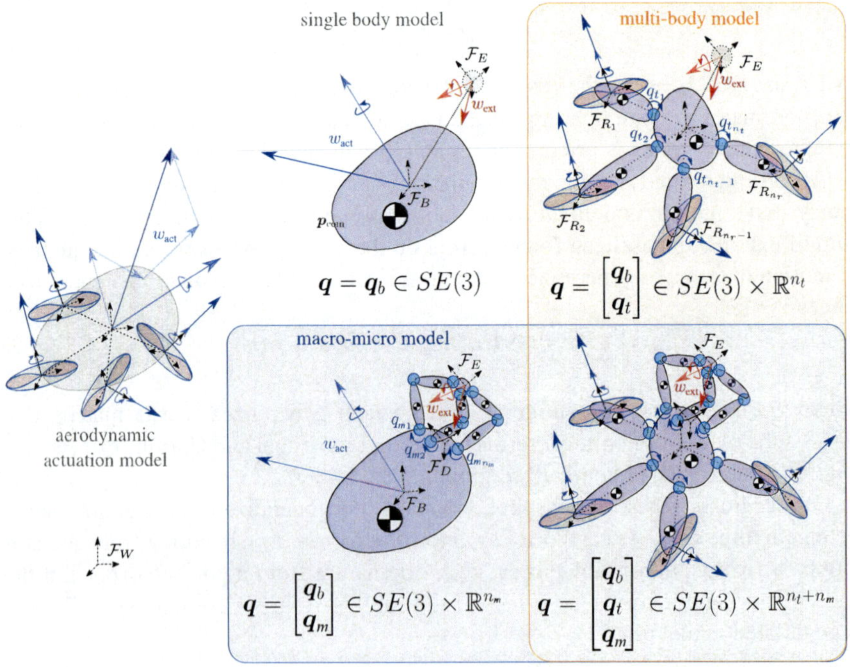

Fig. 3.1 System model overview with generalized coordinates, $\boldsymbol{q}$: A net aerodynamic actuation model (left), a single rigid inertial body (top center), a multi-body system with tilting rotor groups (top right), a macro-micro manipulator (bottom left), and their combination (bottom right)

3.1 Aerodynamic Actuation Model

In this section, we start with the *kinematic description* of a general tilt-rotor system. We further define the *propeller wrench model*, and the *net actuation wrench* based on tilt-able rotor groups, and examine the resulting properties. Methods for evaluating the actuation wrench are presented, including force and torque actuation envelopes. Morphology-based singularity conditions are also identified and discussed.

We start by considering a set of simplifying assumptions:

Assumptions:

- The rotor thrust force and drag torque occur only on the propeller axis, and are related in magnitude by a proportional constant.
- Rotors are able to achieve desired forces with negligible transients.
- The actuation wrenches produced by each rotor group are independent, *i.e.* there is no airflow interference.

Our assumptions consider any off-axis forces generated from aerodynamic effects negligible, and constrain our description of the drag torque magnitude as proportional to the lift force and opposite to the direction of rotor spin. The ability to instantaneously achieve a desired rotor thrust force is related to the idea that the rotors have very small inertia, and therefore the change in force can occur at speeds comparable to our system bandwidth. The accuracy of this assumption requires further study of empirical transients in accelerating and decelerating the rotors, and will differ depending on the implementation of the motor, propeller, and electronic speed controller. Considering each actuation wrench as independent, we move the effects of airflow interference between tilt-able rotor groups into the lumped disturbances acting on the system. Since the aerodynamics of these interfering air flows are difficult to model precisely and would be the subject of computational fluid dynamics, we assume that the interference effect is small compared to the overall net thrust magnitude to greatly simplify our actuation model.

3.1.1 Kinematic Description

Common coordinate frames used for modelling are listed in Table 3.1, and shown in Fig. 3.1 for system models that will be developed throughout this chapter. The inertial (or world) non-accelerating frame is represented by $\mathcal{F}_W$, placed such that the $\boldsymbol{e}_z^W$ axis is opposite to the gravity vector. The floating base frame, $\mathcal{F}_B$, is fixed to the inertial body of the base, and rotates and translates in space relative to $\mathcal{F}_W$. Tilting rotor groups $\mathcal{F}_{R_i}$ are each actuated by a single rotary joint, $q_{t_i} \in \boldsymbol{q}_t$, which defines their transformation relative to $\mathcal{F}_B$. A manipulator base frame, $\mathcal{F}_D$, is rigidly attached to the floating base and is related by a static transformation to $\mathcal{F}_B$. The end-effector frame, $\mathcal{F}_E$, represents the point where an external contact wrench is applied and is

Table 3.1 Main coordinate frames with their origin and principal axes, and their degrees of freedom relative to associated coordinate frames

Frame	Subscript	Definition	DOFs	Rel. to frame
World	W	$\mathcal{F}_W \in \{O_W, \boldsymbol{e}_x^W, \boldsymbol{e}_y^W, \boldsymbol{e}_z^W\}$		
Base	B	$\mathcal{F}_B \in \{O_B, \boldsymbol{e}_x^B, \boldsymbol{e}_y^B, \boldsymbol{e}_z^B\}$	6	$\mathcal{F}_W$
ith rotor group	R_i	$\mathcal{F}_{R_i} \in \{O_{R_i}, \boldsymbol{e}_x^{R_i}, \boldsymbol{e}_y^{R_i}, \boldsymbol{e}_z^{R_i}\}$	1	$\mathcal{F}_B$
Delta arm base	D	$\mathcal{F}_D \in \{O_D, \boldsymbol{e}_x^D, \boldsymbol{e}_y^D, \boldsymbol{e}_z^D\}$	0	$\mathcal{F}_B$
End-effector	E	$\mathcal{F}_E \in \{O_E, \boldsymbol{e}_x^E, \boldsymbol{e}_y^E, \boldsymbol{e}_z^E\}$	0/3	$\mathcal{F}_D$

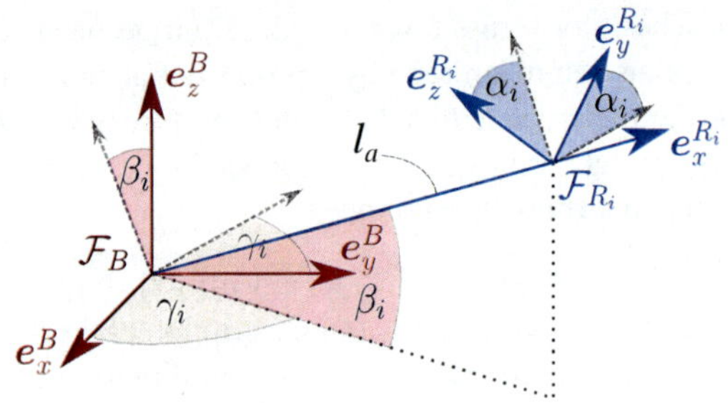

Fig. 3.2 Illustration of transformation between base and rotor group frames

either a static transformation from $\mathcal{F}_B$ in the case of a rigidly mounted manipulator, or defined by actuated manipulator joints, $\boldsymbol{q}_m$, in the case of a macro-micro manipulator model.

We express the transformation of each rotor group frame $\mathcal{F}_{R_i}$ relative to the base frame $\mathcal{F}_B$ as shown in Fig. 3.2 to develop the aerodynamic actuation model. We model the tilt-rotor system as n_t rotor group frames $\mathcal{F}_{R_i}$ positioned around $\mathcal{F}_B$. The rotating axis of each rotor group, $\boldsymbol{e}_x^{R_i}$, is aligned with the axis of the support arm coming out of the main body, and intersects the base frame origin, O_B. This arm of length l_a may be positioned at arbitrary angles γ_i, β_i which respectively define the rotation about the $\boldsymbol{e}_z^B$ axis, and inclination from the $\boldsymbol{e}_z^B$ plane. The actuated tilting joint acts about the $\boldsymbol{e}_x^{R_i}$ axis, controlling the configuration variable $q_{t_i} = \alpha_i$ for $i \in \{1 \ldots n_t\}$ The frame transformation is constructed as

$$\boldsymbol{T}_{BR_i} = \underbrace{\begin{bmatrix} \boldsymbol{R}_z(\gamma_i) & \boldsymbol{0}_{3\times1} \\ \boldsymbol{0}_{1\times3} & 1 \end{bmatrix}}_{\text{rotation about } \boldsymbol{e}_z \text{ by } \gamma_i} \underbrace{\begin{bmatrix} \boldsymbol{R}_y(\beta_i) & \boldsymbol{0}_{3\times1} \\ \boldsymbol{0}_{1\times3} & 1 \end{bmatrix}}_{\text{rotation about } \boldsymbol{e}_y \text{ by } \beta_i} \underbrace{\begin{bmatrix} \mathbb{1}_3 & l_a\boldsymbol{e}_x \\ \boldsymbol{0}_{1\times3} & 1 \end{bmatrix}}_{\text{translation along } \boldsymbol{e}_x \text{ by } l_a} \underbrace{\begin{bmatrix} \boldsymbol{R}_x(\alpha_i) & \boldsymbol{0}_{3\times1} \\ \boldsymbol{0}_{1\times3} & 1 \end{bmatrix}}_{\text{rotation about } \boldsymbol{e}_x \text{ by } \alpha_i}, \tag{3.4}$$

for all $i \in \{1, ..., n_t\}$.

Configuration variables are represented in vector form as

$$\boldsymbol{\gamma} = \begin{bmatrix} \gamma_1 \\ \vdots \\ \gamma_{n_t} \end{bmatrix} \quad \boldsymbol{\beta} = \begin{bmatrix} \beta_1 \\ \vdots \\ \beta_{n_t} \end{bmatrix} \quad \boldsymbol{\alpha} = \begin{bmatrix} \alpha_1 \\ \vdots \\ \alpha_{n_t} \end{bmatrix}. \tag{3.5}$$

For the tilt-rotor models we explore in this book, $\boldsymbol{\gamma}$ and $\boldsymbol{\beta}$ are fixed variables, and $\boldsymbol{\alpha}(t)$ is a time-dependent controllable variable.

3.1.2 Propeller Wrench Model

Thrust forces along and drag torques about the rotational axis of a physical air screw are governed by rotary-wing aerodynamics [1]. We take the simplifying assumption that the force and torque of the propeller are instantaneously proportional to the square of the propeller's angular velocity ω_{p_i} for the ith propeller where $i \in \{1, ..., n_r\}$. These proportional coefficients include the effects of chord length and pitch over the length of the blade, as well as air density which we assume to be constant.

We follow the common modelling approximation that the force and torque produced by a rotor can be described as

$$f_{r_i} = c_f \omega_{p_i}^2 \tag{3.6a}$$

$$\tau_{r_i} = -s_{r_i} c_f c_d \omega_{p_i}^2 \tag{3.6b}$$

where c_f is the *rotor thrust coefficient* and c_d is the *rotor drag coefficient*, which in this formulation also measures the thrust efficiency relative to drag torque. Both c_f and c_d are positive constant scalars associated with a particular propeller.

Since we can write τ_{r_i} directly in terms of f_{r_i} as $\tau_{r_i} = -s_{r_i} c_d f_{r_i}$, the entire actuation wrench at a single propeller in its rotor group frame can be described as a function of this scalar value:

$$\boldsymbol{w}_{R_i} = \begin{bmatrix} \boldsymbol{f}_{R_i} \\ \boldsymbol{\tau}_{R_i} \end{bmatrix} = f_{r_i} \begin{bmatrix} 0 \\ 0 \\ 1 \\ 0 \\ 0 \\ -s_{r_i} c_d \end{bmatrix} \quad \forall i \in \{1, ..., n_r\}, \tag{3.7}$$

Each propeller generates a pure thrust force along $\boldsymbol{e}_z^{R_i}$ of magnitude f_{r_i} and a pure drag torque about $\boldsymbol{e}_z^{R_i}$ of $-s_{r_i} c_d f_{r_i}$ (opposite to the direction of propeller spin, s_{r_i}).

Rotor forces are subject to actuation limits due to physical limitations of the electro-mechanical rotor-propeller subsystem, $f_{r_{\min}}$ and $f_{r_{\max}}$ where a negative value for $f_{r_{\min}}$ corresponds to a propeller spin inversion, producing negative thrust and inverting the direction of drag torque.

$$f_{r_{\min}} \leq f_{r_i} \leq f_{r_{\max}} \quad \forall i \in \{1, ..., n_r\} \tag{3.8}$$

3.1.3 Double Propeller Group

We further consider double rotor-propeller elements located at $\mathcal{F}_{R_i}$. Two co-axially stacked propellers compared to a single rotor group have the theoretical advantage of doubling the thrust force while only marginally increasing the weight and size of the entire system.

Airflow interference between the two propellers results in some loss of efficiency, and is primarily dependent on the propeller spacing. Rotary-wing aerial vehicles work on the principle of air screws pushing air downwards to generate lift, and with collinear rotor groups, the lower rotor group's air inflow has already been accelerated by the upper rotor, as depicted in Fig. 3.3.

State-of-the-art research has investigated the influence of collinear rotor groups [2], and determined an empirical relationship between rotor plane spacing and the loss of thrust compared to two non-interfering rotors. This relationship indicates that spacing the propeller planes apart by at least $0.3\times$ the rotor diameter results in minimal thrust losses, but still in the rage of 20%, with almost the entire loss occurring at the lower rotor (40%). In addition, the further apart the rotor groups are spaced, the larger the rotor group's rotational inertia becomes about the tilting axis, and the larger the occupied volume of the entire system becomes.

For co-rotating propeller groups (same spin direction), the net drag torque theoretically remains a multiple of the net thrust force, despite 40% efficiency loss of the

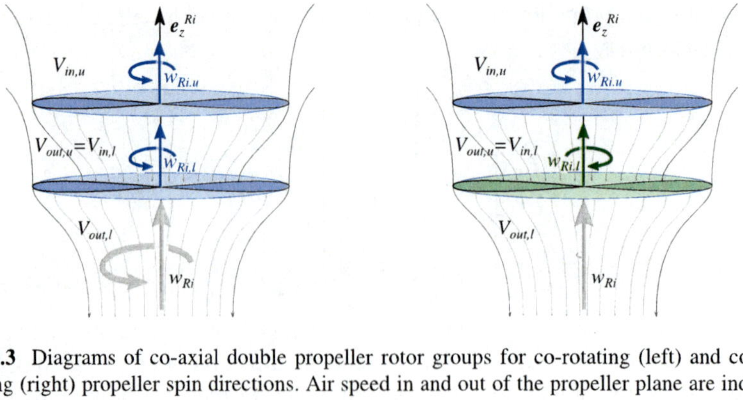

Fig. 3.3 Diagrams of co-axial double propeller rotor groups for co-rotating (left) and counter-rotating (right) propeller spin directions. Air speed in and out of the propeller plane are indicated with $V_{\text{in/out}}$

lower rotor. The net effect at the ith rotor group can be approximated as a single propeller:

$$\boldsymbol{w}_{R_i} = c_f \omega_{p_i u}^2 \begin{bmatrix} 0 \\ 0 \\ 1 \\ 0 \\ 0 \\ -s_{r_i} c_d \end{bmatrix} + 0.6\, c_f \omega_{p_i l}^2 \begin{bmatrix} 0 \\ 0 \\ 1 \\ 0 \\ 0 \\ -s_{r_i} c_d \end{bmatrix} = \underbrace{c_f (\omega_{p_i u}^2 + 0.6\, \omega_{p_i l}^2)}_{f_{r_i}} \begin{bmatrix} 0 \\ 0 \\ 1 \\ 0 \\ 0 \\ -s_{r_i} c_d \end{bmatrix}, \tag{3.9}$$

where $\omega_{p_i u}$ and $\omega_{p_i l}$ are the upper and lower propeller angular velocities.

For counter-rotating groups, the opposing drag torques result in a model that cannot be simplified back to a single force variable:

$$\boldsymbol{w}_{R_i} = \underbrace{c_f \omega_{p_i u}^2}_{f_{r_{i,u}}} \begin{bmatrix} 0 \\ 0 \\ 1 \\ 0 \\ 0 \\ -s_{r_{i,u}} c_d \end{bmatrix} + \underbrace{0.6\, c_f \omega_{p_i l}^2}_{f_{r_{i,l}}} \begin{bmatrix} 0 \\ 0 \\ 1 \\ 0 \\ 0 \\ -s_{r_{i,l}} c_d \end{bmatrix}. \tag{3.10}$$

In the case of counter-rotating co-axial rotor groups which are used in the contents of this work, we ignore the complexities of airflow dynamics involved in a double rotor group, and empirically model the net effect in one of two ways. We either consider the two propellers as independent, non-interfering generators of rotor thrust and drag torque, with an empirically applied penalty to the lower rotor (reduced c_f) to simulate the reduced efficiency from empirically gathered data, or model them as a single rotor group with pure thrust and negligible drag torque ($c_d \approx 0$).

3.1.4 Net Aerodynamic Actuation Wrench

The net actuation wrench is computed as the sum of all rotor group wrenches in $\mathcal{F}_B$. The wrench transformation between $\mathcal{F}_B$ and $\mathcal{F}_{R_i}$ is parameterized with configuration variables $\{\alpha_i, \beta_i, \gamma_i\}$, and transformed as follows:

$${}^B\boldsymbol{w}_{R_i} = \boldsymbol{Ad}^\top_{\boldsymbol{T}^{-1}_{BR_i}}(\alpha_i, \beta_i, \gamma_i)\, \boldsymbol{w}_{R_i} = \begin{bmatrix} \boldsymbol{R}_{BR_i} \boldsymbol{f}_{R_i} \\ \boldsymbol{p}_{BR_i} \times \boldsymbol{R}_{BR_i} \boldsymbol{f}_{R_i} + \boldsymbol{R}_{BR_i} \boldsymbol{\tau}_{R_i} \end{bmatrix} = \boldsymbol{a}_i f_{r_i} \tag{3.11}$$

where $\boldsymbol{a}_i$ is a configuration-based vector representing ith rotor wrench in $\mathcal{F}_B$, when multiplied by the rotor thrust force. Since the effect of each of n_r rotor groups is independent, we can compute the net force by combining the columns to form the *instantaneous actuator allocation matrix* $\boldsymbol{A}_{\text{act}} \in \mathbb{R}^{6 \times n_r}$.

$$
{}^{B}\boldsymbol{w}_{\mathrm{act}} = \sum_{i=1}^{n_r} {}^{B}\boldsymbol{w}_{R_i} = \begin{bmatrix}\boldsymbol{a}_1 \cdots \boldsymbol{a}_{n_r}\end{bmatrix} \underbrace{\begin{bmatrix} f_{r_1} \\ \vdots \\ f_{r_{n_r}} \end{bmatrix}}_{\boldsymbol{f}_r} = \boldsymbol{A}_{\mathrm{act}} \boldsymbol{f}_r \tag{3.12}
$$

The term 'instantaneous' refers to our assumptions on instantaneous rotor thrust dynamics. For tilt-rotor platforms, $\boldsymbol{\alpha}(t)$ values change over time with non-negligible dynamics. As a result, $\boldsymbol{A}_{\mathrm{act}}(\boldsymbol{\alpha}(t))$ also varies with time as a function of the tilt-able coordinates.

3.1.5 Tilting Joint Model

With tilt-able rotor groups, the actuator allocation matrix $\boldsymbol{A}_{\mathrm{act}}$ is not constant, and is a function of the current tilt angles, $\boldsymbol{\alpha}$. Commanding the motion of a larger inertial body, these tilting rotary joints are subject to dynamic actuation constraints. In turn, the change in the net aerodynamic actuation wrench of a system is also subject to constraints.

We express the time evolution of an actuated rotary joint for variable α_i as a simplified first-order damped system (as previously done in [3]):

$$
\dot{\alpha}_i = \frac{1}{\tau_\alpha}\left(\alpha_i^* - \alpha_i\right) \tag{3.13}
$$

where α_i^* and α_i are respectively the reference and measured joint angles, and τ_α is a positive scalar representing the time constant of the rotary motion. This model includes dynamic effects and some physical limitations.

A given tilt-rotor joint can turn in either direction with no hard joint constraints, but is subject to a velocity limit, $\dot{\alpha}_{\max}$:

$$
-\dot{\alpha}_{\max} \le \dot{\alpha}_i \le \dot{\alpha}_{\max} \quad \forall i \in \{1, \ldots, n_r\}. \tag{3.14}
$$

In reality, the motor limits relate to current or torque, $\tau_{\alpha,\max}$, which will be a useful constraint formulation for whole-body control of the tilt-rotor system:

$$
-\tau_{\alpha,\max} \le \tau_{\alpha_i} \le \tau_{\alpha,\max} \quad \forall i \in \{1, \ldots, n_r\}. \tag{3.15}
$$

3.1.6 Force and Torque Actuation Volumes

Volumes corresponding to the maximum reachable net actuation forces (f_{vol}) and torques (τ_{vol}) of the system expressed in $\mathcal{F}_B$ are a valuable representation for evaluating a system's omnidirectional actuation capabilities. We refer to these 3D representations as *force* and *torque envelopes* or *actuation volumes*.

Points on the surface of the 3D force volume are found by choosing a dense array of evenly spaced direction vectors, and for each vector $\boldsymbol{d} \in \mathbb{R}^3$, finding the maximum force magnitude, f^d, with zero torque that does not exceed actuation limits.

$$\begin{aligned} \max_{\alpha, f_r} \quad & f^d \\ \text{s.t.} \quad & f^d \begin{bmatrix} \boldsymbol{d} \\ \boldsymbol{0} \end{bmatrix} = \boldsymbol{A}_{\text{act}}(\boldsymbol{\alpha}) \boldsymbol{f}_r \\ & f_{r_{\min}} \leq f_{r_i} \leq f_{r_{\min}} \quad \forall i \in 1 \ldots n_r \end{aligned} \tag{3.16}$$

The force volume is nominally in the absence of torque, when computed at the COM of the system. Where there is an offset in the center of control and the COM, the torque required to statically maintain a given orientation should be applied during the search for actuator saturation. This volume will then change depending on the orientation, finding its maximum where the COM aligns with the center of control along the gravitational axis.

$$\begin{aligned} \max_{\alpha, f_r} \quad & \tau^d \\ \text{s.t.} \quad & \tau^d \begin{bmatrix} {}^B\boldsymbol{g} \\ \boldsymbol{d} \end{bmatrix} = \boldsymbol{A}_{\text{act}}(\boldsymbol{\alpha}) \boldsymbol{f}_r \\ & f_{r_{\min}} \leq f_{r_i} \leq f_{r_{\min}} \quad \forall i \subset 1 \ldots n_r \end{aligned} \tag{3.17}$$

Similarly for the torque envelope, A volume should be computed with a gravity-compensating force ${}^B\boldsymbol{g}$ required to maintain static hover for any common flight configuration. Design symmetry can be an advantage in having fewer envelopes to consider. A conservative volume can also be computed, taking the minimum intersection of volumes for all hover orientations.

Meaningful and comparable properties from force and torque volumes are the minimum and maximum magnitude (f_{max}, f_{min}, τ_{max}, τ_{min}), the mean surface point radius (f_{mean}, τ_{mean}), and the total volume within the defined surface (f_{vol}, τ_{vol}). These values can be used to evaluate and compare dynamic capabilities.

It is worth noting that the envelopes described above do not capture delays in wrench rate tracking due to the slower dynamics of $\boldsymbol{\alpha}$, though the actuation wrench rate is an important property of platform actuation. This property is difficult to study in a 3D plotted envelope, since it changes depending on the configuration of tilt-rotor joints, and comparison requires imposing limits on the change in rotor thrust forces, which at the moment are assumed to be instantaneous.

3.1.7 Singularity Conditions

Studying the actuator allocation, we encounter two types of singularity cases: an instantaneous rank reduction of the instantaneous allocation matrix, and critical condition of undefined tilt angles.

Instantaneous Rank Reduction

A singular condition occurs when the tilt-rotor coordinates cause a rank reduction in $\boldsymbol{A}_{\text{act}}(\boldsymbol{\alpha})$. At time t_s, $\boldsymbol{\alpha}(t_s)$ align the propeller axes such that instantaneous controllability of select forces and torques is lost, and must be regained by re-orientation of the tilt-able rotor groups. We note that the system can still meet the definition of fully actuated in Eq. (2.9) if the net actuation wrench can change in any direction, which is achieved by tilt-rotor motion.

Mathematically, the scenarios where these singularities occur can be found by studying the condition number of $\boldsymbol{A}_{\text{act}}$, defined as

$$\kappa(\boldsymbol{A}_{\text{act}}(\boldsymbol{\alpha})) = \frac{\sigma_{\max}(\boldsymbol{A}_{\text{act}}(\boldsymbol{\alpha}))}{\sigma_{\min}(\boldsymbol{A}_{\text{act}}(\boldsymbol{\alpha}))} \tag{3.18}$$

where $\sigma_{\max}$ and $\sigma_{\min}$ are the maximum and minimum singular values of $\boldsymbol{A}_{\text{act}}(\boldsymbol{\alpha})$. Since a large number of configurations are possible, we consider the selection of $\boldsymbol{\alpha}$ that corresponds with the minimum amount of parasitic forces, as used for the computation of force and torque envelopes.

We generate *condition number volumes* by plotting the log of $\kappa(\boldsymbol{A}_{\text{act}}(\boldsymbol{\alpha}))$ over a force envelope of magnitude mg with an additional static in two hover force conditions: along the $\boldsymbol{e}_y^B$ and $\boldsymbol{e}_z^B$ axes. The singular configurations are apparent in Fig. 3.4, where the condition number asymptotically extends along the $\boldsymbol{e}_z^B$ axis and $\boldsymbol{e}_z^B$ plane

These conditions are present when the desired force aligns with the $\boldsymbol{e}_z^B$ axis (rank is reduced to 4, no instantaneous force exertion along $\boldsymbol{e}_x^B$ or $\boldsymbol{e}_y^B$ axes is possible), and

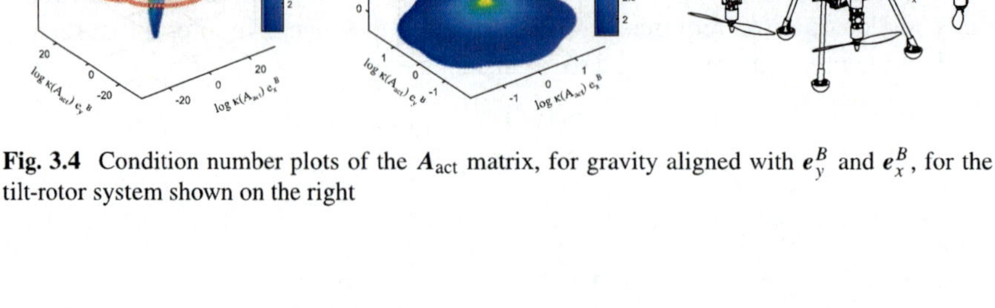

Fig. 3.4 Condition number plots of the $\boldsymbol{A}_{\text{act}}$ matrix, for gravity aligned with $\boldsymbol{e}_y^B$ and $\boldsymbol{e}_x^B$, for the tilt-rotor system shown on the right

when the desired force lies on the $\boldsymbol{e}_z^B$ plane (rank is reduced to 5, no instantaneous force exertion is possible along $\boldsymbol{e}_z^B$).

Kinematic Singularity

A second singular condition occurs when a rotor thrust f_{r_i} cannot contribute to the desired net actuation wrench ${}^B\boldsymbol{w}_{\text{act}}$, which leads to the corresponding tilt angle α_i not being uniquely defined. This case occurs when the desired net force aligns with the arm axis $\boldsymbol{e}_x^{R_i}$, and tilt angle dynamic limitations do not appear explicitly in the instantaneous actuator allocation model. When the desired force vector passes through this axis directly, tracking the force requires infinite $\dot{\alpha}_i$ as $dt \rightarrow 0$.

To expose the tilt-rotor angular velocities, $\dot{\boldsymbol{\alpha}}$, we differentiate the actuation ${}^B\boldsymbol{w}_{\text{act}}$ with respect to time via tilt coordinates $\boldsymbol{\alpha}$ and rotor forces $\boldsymbol{f}_r$, and form the *differential actuator allocation matrix*.

Starting from Eq. (3.12),

$$\frac{d}{dt}({}^B\boldsymbol{w}_{\text{act}}) = \frac{d}{dt}(\boldsymbol{A}_{\text{act}}(\boldsymbol{\alpha})\boldsymbol{f}_r \tag{3.19}$$

$${}^B\dot{\boldsymbol{w}}_{\text{act}} = \underbrace{\frac{\partial \boldsymbol{A}_{\text{act}}(\boldsymbol{\alpha})\boldsymbol{f}_r}{\partial \boldsymbol{f}_r}}_{\boldsymbol{A}_{\text{act}}(\boldsymbol{\alpha})}\underbrace{\frac{\partial \boldsymbol{f}_r}{\partial t}}_{\dot{\boldsymbol{f}}_r} + \frac{\partial \boldsymbol{A}_{\text{act}}(\boldsymbol{\alpha})\boldsymbol{f}_r}{\partial \boldsymbol{\alpha}}\underbrace{\frac{\partial \boldsymbol{\alpha}}{\partial t}}_{\dot{\boldsymbol{\alpha}}} \tag{3.20}$$

Now, we reformulate an allocation matrix in terms of force and tilt-angle rates:

$${}^B\dot{\boldsymbol{w}}_{\text{act}} = \left[\boldsymbol{A}_{\text{act}}(\boldsymbol{\alpha}) \;\; \frac{\partial \boldsymbol{A}_{\text{act}}(\boldsymbol{\alpha})}{\partial \boldsymbol{\alpha}}\boldsymbol{f}_r\right]\begin{bmatrix}\dot{\boldsymbol{f}}_r \\ \dot{\boldsymbol{\alpha}}\end{bmatrix} \tag{3.21}$$

From expression Eq. (3.21) we see that any value of $\dot{\alpha}_i$ satisfies the equation when the corresponding f_{r_i} term is zero.

3.2 Omnidirectional Tilt-Rotor Model

For the system dynamics of an omnidirectional flying robot, we start by deriving the simplest model of the tilt-rotor system as a **single rigid body**, then further develop the preliminaries for the tilt-rotor model as a **multi-body system**, extending the coordinates to include tilt-rotor joints. A **morphology optimization** problem is then explored to choose a tilt-rotor morphology for the remainder of this work.

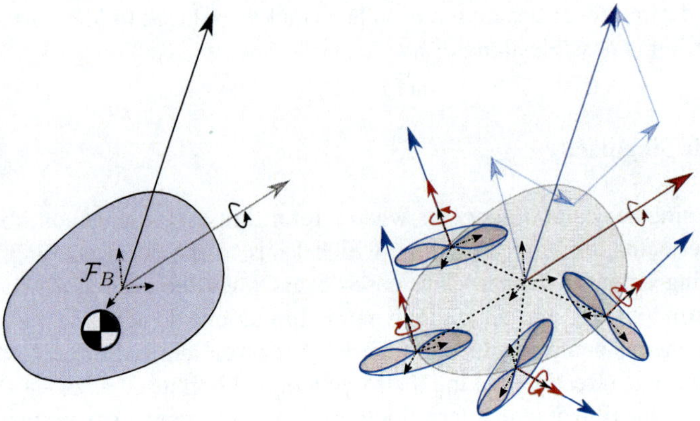

Fig. 3.5 Left: Single rigid body flying base in free flight. **Right:** Collective actuation wrench represented as the sum of all individual rotor wrenches expressed in the base frame

3.2.1 Single Body Dynamic Model

For the most simplified and general treatment of a fully actuated flying base as a dynamic system, we model the entire robot as a single rigid body. Figure 3.5 shows the model of a single rigid body in free flight, with its collective actuation wrench. We select coordinates $\boldsymbol{q}$, define the dynamic terms $\boldsymbol{\mathcal{M}}$, $\boldsymbol{c}$ and $\boldsymbol{g}$ from Eq. (3.2) for a single floating rigid body, and the actuation model as the sum of rotor forces occurring in the base frame. We use the following assumptions to define the simplified system model.

Assumptions:

- The entire platform can be modelled as a single rigid body.
- The dynamics of tilting rotor groups and propellers are considered negligible.
- The aerodynamic actuation forces can be modelled as a single wrench acting at the platform base.

A single rigid body that moves arbitrarily in 3D space has 3 translational and 3 rotational DOFs, which we represent with our choice of generalized coordinates $\boldsymbol{q}$ for the system. The rigid body system has a body-fixed base frame, $\mathcal{F}_B$. Our translational coordinate representation is the position of the base frame origin expressed in the $\mathcal{F}_W$ (see Eq. (3.22)). We represent the system's orientation on the SO(3) manifold to avoid singularities and ambiguity associated with other representations such as Euler angles (see Eq. (3.23)). The resulting coordinates for the floating base are then

$$\boldsymbol{q}_{b_p} = {}^{W}\boldsymbol{p}_B \in \mathbb{R}^3 \tag{3.22}$$

$$\boldsymbol{q}_{b_R} = \boldsymbol{R}_{WB} \in SO(3) \tag{3.23}$$

$$\boldsymbol{q} = \boldsymbol{q}_b = \{\boldsymbol{q}_{b_p}, \boldsymbol{q}_{b_R}\} \in \underbrace{\mathbb{R}^3 \times SO(3)}_{SE(3)} \tag{3.24}$$

where SE(3) is the special Euclidean group used to uniquely represent 3D rigid body transformations.

In deriving the system's equations of motion, the Lagrange formulation is complicated by our choice of orientation coordinates on SE(3), since the time derivative of the coordinates is not directly the generalized velocity vector. As such, the Newton-Euler formulation is more practical for this derivation, following the conservation of momentum law.

The system is modelled as an inertial body with mass $m \in \mathbb{R}$ and inertia $\mathcal{I}_B \in \mathbb{R}^{3\times3}$ expressed in $\mathcal{F}_B$, subject to gravitational body forces with its COM located at ${}^{B}\boldsymbol{p}_{\otimes}$. It can be actuated by an arbitrary control wrench vector exerted at the *center of control* which coincides with the base frame origin O_B. General aerodynamic disturbance forces may act on the system, caused by gusts of wind and aerodynamic factors, and are represented by their net effect at $\mathcal{F}_B$. A general contact wrench may act at a certain location $\mathcal{F}_E$ on the system.

We choose to formulate the Newton-Euler equations of motion in the floating base frame $\mathcal{F}_B$, since this is the frame in which our actuation will take place, and where our inertial sensors are located. The generalized velocity vector consists of the linear velocity of O_B with respect to O_W rotated into the base frame, $\boldsymbol{v}_B$, and the angular velocity of $\mathcal{F}_B$ with respect to $\mathcal{F}_W$ expressed in the rotating base frame, $\boldsymbol{\omega}_B$.

$$\widetilde{\boldsymbol{u}} = \begin{bmatrix} \boldsymbol{v}_B \\ \boldsymbol{\omega}_B \end{bmatrix} \in \mathbb{R}^6 \tag{3.25}$$

Considering the above forces on the system, we express the change in linear momentum in the inertial frame as

$$ {}^{W}\boldsymbol{f} = \frac{d}{dt}(m\boldsymbol{R}_{WB}\boldsymbol{v}_B) = \boldsymbol{R}_{WB}m\dot{\boldsymbol{v}}_B + \dot{\boldsymbol{R}}_{WB}m\boldsymbol{v}_B \tag{3.26}$$

and rotate the resulting expression into the base frame, using the relation that $\hat{\boldsymbol{\omega}}_B = \boldsymbol{R}_{WB}^{\top}\dot{\boldsymbol{R}}_{WB}$ as derived in Sect. A.1,

$$ {}^{B}\boldsymbol{f} = \boldsymbol{R}_{WB}^{\top}\, {}^{W}\boldsymbol{f} = m(\dot{\boldsymbol{v}}_B + {}^{B}\boldsymbol{g}) + \boldsymbol{\omega}_B \times m\boldsymbol{v}_B. \tag{3.27}$$

In the above equations, we include the body acceleration due to gravity. The gravitational wrench can be modelled as the system's mass times a pure linear acceleration g in $\boldsymbol{e}_z^W$ acting at the COM, rotated into the base frame as ${}^{B}\boldsymbol{g} = \boldsymbol{R}_{WB}^{\top}[0\ 0\ g]^{\top}$.

For the angular dynamics, since the base frame is not necessarily coincident with the center of mass, we can use the parallel axis theorem to express the constant inertia matrix in $\mathcal{F}_B$ as $\mathcal{I}_B = \mathcal{I}_{\circledast} - m[\,{}^B\hat{\boldsymbol{p}}_{\circledast}][\,{}^B\hat{\boldsymbol{p}}_{\circledast}]$, or simply determine the inertia in $\mathcal{F}_B$ using a detailed CAD model. Similarly to the linear derivation, we express the change in angular momentum as

$$ {}^B\boldsymbol{\tau} = \mathcal{I}_B\dot{\boldsymbol{\omega}}_B + \boldsymbol{\omega}_B \times \mathcal{I}_B\boldsymbol{\omega}_B + {}^B\boldsymbol{p}_{\circledast} \times {}^B\boldsymbol{f}, \tag{3.28} $$

where the final term accounts for a center of mass offset from the base frame.

Combining Eqs. (3.27) and (3.28) with forces acting at the base frame, the resulting equations of motion in $\mathcal{F}_B$ are

$$ \underbrace{\begin{bmatrix} \sum {}^B\boldsymbol{f} \\ \sum {}^B\boldsymbol{\tau} \end{bmatrix}}_{\substack{\text{net force and torque}\\ \text{acting on base frame}}} = \underbrace{\begin{bmatrix} m\mathbb{1}_3 & \boldsymbol{0}_{3\times 3} \\ m\,{}^B\hat{\boldsymbol{p}}_{\circledast} & \mathcal{I}_B \end{bmatrix}}_{\substack{\text{generalized inertia}\\ \text{matrix } \boldsymbol{M}}} \underbrace{\begin{bmatrix} \dot{\boldsymbol{v}}_B \\ \dot{\boldsymbol{\omega}}_B \end{bmatrix}}_{\substack{\text{base frame}\\ \text{acceleration } \dot{\boldsymbol{\nu}}_B}} $$

$$ + \underbrace{\begin{bmatrix} m\hat{\boldsymbol{\omega}}_B & \boldsymbol{0} \\ m\,{}^B\hat{\boldsymbol{p}}_{\circledast}\hat{\boldsymbol{\omega}}_B & \boldsymbol{\omega}_B \times \mathcal{I} \end{bmatrix}}_{\substack{\text{centrifugal and}\\ \text{Coriolis terms } \boldsymbol{C}}} \underbrace{\begin{bmatrix} \boldsymbol{v}_B \\ \boldsymbol{\omega}_B \end{bmatrix}}_{\substack{\text{base frame}\\ \text{twist } \boldsymbol{\nu}_B}} + \underbrace{\begin{bmatrix} m\,{}^B\boldsymbol{g} \\ {}^B\boldsymbol{p}_{\circledast} \times m\,{}^B\boldsymbol{g} \end{bmatrix}}_{\text{gravitational terms } \boldsymbol{g}} $$

$$ = \underbrace{\begin{bmatrix} {}^B\boldsymbol{f}_{\text{act}} \\ {}^B\boldsymbol{\tau}_{\text{act}} \end{bmatrix}}_{\substack{\text{aerodynamic}\\ \text{actuation wrench}}} + \underbrace{\begin{bmatrix} {}^B\boldsymbol{f}_{\Delta} \\ {}^B\boldsymbol{\tau}_{\Delta} \end{bmatrix}}_{\substack{\text{disturbances and}\\ \text{model error}}} + \underbrace{\begin{bmatrix} {}^B\boldsymbol{f}_{c} \\ {}^B\boldsymbol{\tau}_{c} \end{bmatrix}}_{\substack{\text{contact}\\ \text{forces}}}, \tag{3.29} $$

where external forces acting on the system which induce a change in linear and angular momentum are the actuation wrench, a contact wrench, and other lumped disturbances. We note from our derivations that the inertial matrix $\boldsymbol{M}$ is constant since we express it in the base frame, matrix $\boldsymbol{C}$ containing fictitious forces is a function of the current twist and orientation, and the gravitational vector $\boldsymbol{g}$ is a function of the current base orientation.

3.2.2 Multi-body Dynamic Model

In this section, we extend the floating base model to include tilt-able rotor groups as separate bodies for a more accurate dynamic model of a tilt-rotor aerial vehicle. The multi-body formulation is addressed in a preliminary sense, with the aim to aid in future full-body controller development. This model removes the assumption that tilting rotor group dynamics are negligible, and is accompanied by the new assumption that the dynamics of tilt motors are independent of the rotational speed

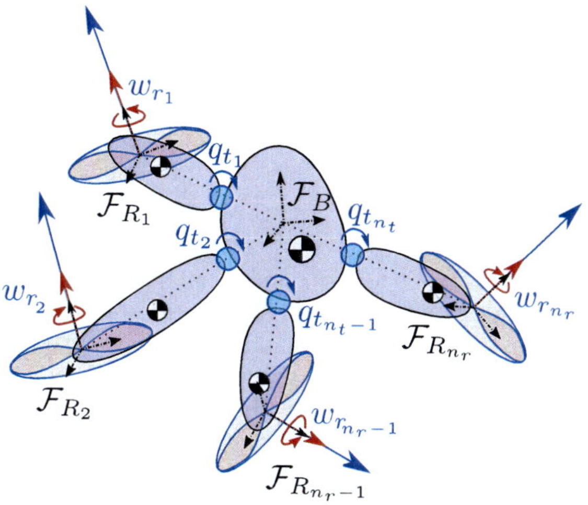

Fig. 3.6 Generic model of a multi-body tilt-rotor aerial system

of rotors. The new assumption allows us to neglect the rigid body dynamics of the propellers, pushing these relatively small effects into the lumped disturbance term.

Removing the assumption that the system is a single rigid body, we introduce the dynamics of a rigid multi-body system connected by actuated joints, as seen in Fig. 3.6. Each of the new rigid bodies is connected to the flying base by an open-chain, and has its own inertial properties ($\mathcal{I}$, m, $\boldsymbol{p}_{\otimes}$). We extend the state of the system to include actuated tilt variables, $\boldsymbol{q}_t$. In the case of the tilt-rotor model, the new rigid bodies are rotor groups connected by independently actuated rotary joints, which control the variables $q_{t_i} = \alpha_i(t) \forall i \in \{1, .., n_t\}$. The system coordinates and generalized velocity vector are now

$$\boldsymbol{q} = \begin{bmatrix} \boldsymbol{q}_b \\ \boldsymbol{q}_t \end{bmatrix} = \begin{bmatrix} {}^W\boldsymbol{p}_B \\ \boldsymbol{R}_{WB} \\ \boldsymbol{\alpha} \end{bmatrix} \in SE(3) \times \mathbb{R}^{n_t} \qquad \widetilde{\boldsymbol{u}} = \begin{bmatrix} \boldsymbol{v}_B \\ \boldsymbol{\omega}_B \\ \dot{\boldsymbol{\alpha}} \end{bmatrix} \in \mathbb{R}^{6+n_t}. \tag{3.30}$$

Similarly, the generalized actuation forces of the system are now extended to include joint torques associated with the tilt-arm velocities,

$$\widetilde{\boldsymbol{\tau}}_{\text{act}} = \begin{bmatrix} {}^B\boldsymbol{f}_{\text{act}} \\ {}^B\boldsymbol{\tau}_{\text{act}} \\ \boldsymbol{\tau}_{\alpha} \end{bmatrix} \in \mathbb{R}^{6+n_t}. \tag{3.31}$$

We use the projected Newton-Euler method to derive the dynamic equations of the multi-body system. This recursive algorithm projects the motion of the system with a forward pass through the kinematic chain, and joint torques in a backward pass projecting the body dynamics onto each joint. Through a forward pass of the dynamics, we arrive at an expression for the linear and angular acceleration, and the angular velocity of each rotor group in its own frame:

$$ {}^{R_i}\boldsymbol{\omega}_{R_i} = {}^{R_i}\boldsymbol{\omega}_B + \boldsymbol{e}_x \dot{\alpha}_i \tag{3.32} $$

$$ {}^{R_i}\dot{\boldsymbol{\omega}}_{R_i} = {}^{R_i}\dot{\boldsymbol{\omega}}_B + \boldsymbol{e}_x \ddot{\alpha}_i + {}^{R_i}\boldsymbol{\omega}_B \times \dot{\alpha}_i \tag{3.33} $$

$$ {}^{R_i}\dot{\boldsymbol{v}}_{R_i} = \boldsymbol{R}_{BR_i}({}^B\dot{\boldsymbol{v}}_B + {}^B\dot{\boldsymbol{\omega}}_B \times {}^B\boldsymbol{p}_{R_i} + {}^B\boldsymbol{\omega}_B \times {}^B\boldsymbol{\omega}_B \times {}^B\boldsymbol{p}_{R_i}) \tag{3.34} $$

where ${}^{R_i}\boldsymbol{\omega}_B = \boldsymbol{R}_{BR_i}^\top\, {}^B\boldsymbol{\omega}_B$ and ${}^{R_i}\dot{\boldsymbol{\omega}}_B = \boldsymbol{R}_{BR_i}^\top\, {}^B\dot{\boldsymbol{\omega}}_B$.

A backward pass of the PNE algorithm sums the total wrench acting at each tilt joint as a result of the dynamics of its open chain child links. Since all tilt-rotor groups are considered as single child links with the base as their parent, and are purely rotating, the sum of forces and moments due to system dynamics in each rotor frame are

$$ \sum {}^{R_i}\boldsymbol{f}_{R_i} = m_{R_i}({}^{R_i}\dot{\boldsymbol{\omega}}_{R_i} \times {}^{R_i}\boldsymbol{p}_{\otimes,R_i} + {}^{R_i}\boldsymbol{\omega}_{R_i} \times {}^{R_i}\boldsymbol{\omega}_{R_i} \times {}^{R_i}\boldsymbol{p}_{\otimes,R_i}) \tag{3.35} $$

$$ \sum {}^{R_i}\boldsymbol{\tau}_{R_i} = {}^{R_i}\mathcal{I}_{OR_i}\, {}^{R_i}\dot{\boldsymbol{\omega}}_{R_i} + {}^{R_i}\boldsymbol{\omega}_{R_i} \times ({}^{R_i}\mathcal{I}_{OR_i}\, {}^{R_i}\boldsymbol{\omega}_{R_i}), \tag{3.36} $$

and adding rotor actuation forces, we have

$$ {}^{R_i}\boldsymbol{f}_{R_i} = \sum {}^{R_i}\boldsymbol{f}_{R_i} + f_{r_i}\boldsymbol{e}_z \quad {}^{R_i}\boldsymbol{\tau}_{R_i} = \sum {}^{R_i}\boldsymbol{\tau}_{R_i} - s_{r_i} c_d f_{r_i} \boldsymbol{e}_z. \tag{3.37} $$

For each tilt-rotor group, we express the associated joint torque as the projection of the moment of the tilt-rotor body exerted on the base onto the joint axis:

$$ \tau_{\alpha_i} = {}^{R_i}\boldsymbol{\tau}_{R_i}^\top \boldsymbol{e}_x. \tag{3.38} $$

The backward recursive step is then computed for the floating base, summing the transformed dynamic wrench at each joint to the equations of motion derived in Eq. (3.29). Regrouping the terms for the extended state vector, we arrive at new expressions for $\mathcal{M}$ (multiples of $\dot{\tilde{\boldsymbol{u}}}$), $\boldsymbol{C}$ (multiples of $\tilde{\boldsymbol{u}}$) and $\boldsymbol{g}$ (remaining terms).

Mass and Inertia for Modelling

The inertial properties of the multi-body system relate generalized force to acceleration, and therefore shape the agility of the flying system. To compare the dynamic properties of different tilt-rotor morphologies, the mass and inertia are parametrically computed based on a simplified system geometry and realistic component masses.

We consider a constant core mass $m_{c,\text{const}}$ (representing a central structure and computation and sensor hardware for complete on-board autonomy), an additional core mass m_a to provide tilt-rotor actuation for each of n_r rotor groups, and a mass for each rotor group m_r. The mass of each rigid arm tube m_t is a function of its length l_a and length-normalized mass $m_{t,\text{norm}}$.

$$m_c = m_{c,\text{const}} + n \cdot m_a \tag{3.39a}$$

$$m_t = m_{t,\text{norm}} \cdot l_a \tag{3.39b}$$

$$m = m_c + n_r(m_r + m_t) \tag{3.39c}$$

The core inertia is modelled as a solid cylinder centered at O_B with radius r_c and height h_c. Rigid tilt arms that connect to propeller groups are modelled as cylindrical tubes, with radii r_1, r_2 and length l_a. For fairness of comparison, we consider that all systems have single rotor groups with origin O_{R_i}, and a tilt axis $\boldsymbol{e}_x^{R_i}$ aligned with the corresponding arm axis. The inertia of each rotor group is modelled as a cylinder of radius r_r and height h_r, with inertial values in $\boldsymbol{e}_y^{R_i}$ and $\boldsymbol{e}_z^{R_i}$ axes averaged to approximate a system independent of tilt. Values are chosen based on components presented in Chap. 5.

$$\mathcal{I}_c = \text{diag}\left(\begin{bmatrix} \frac{1}{12}m_c(3{r_c}^2 + {h_c}^2) \\ \frac{1}{12}m_c(3{r_c}^2 + {h_c}^2) \\ \frac{1}{2}m_c{r_c}^2 \end{bmatrix}\right) \tag{3.40a}$$

$$\mathcal{I}_t = \text{diag}\left(\begin{bmatrix} \frac{1}{2}m_t{r_c}^2) \\ \frac{1}{12}m_t(3({r_1}^2 + {r_2}^2) + L^2) \\ \frac{1}{12}m_t(3({r_1}^2 + {r_2}^2) + L^2) \end{bmatrix}\right) \tag{3.40b}$$

$$\mathcal{I}_r = \text{diag}\left(\begin{bmatrix} \frac{1}{12}m_r(3{r_r}^2 + {h_r}^2) \\ \frac{1}{2}(\frac{1}{12}m_r(3{r_r}^2 + {h_r}^2) + \frac{1}{2}m_r{r_r}^2) \\ \frac{1}{2}(\frac{1}{12}m_r(3{r_r}^2 + {h_r}^2) + \frac{1}{2}m_r{r_r}^2) \end{bmatrix}\right) \tag{3.40c}$$

System inertia is then computed using the parallel axis theorem to express all inertia in $\mathcal{F}_B$:

$$\boldsymbol{p}_t = \left[\tfrac{l_a}{2}\ 0\ 0\right]^\top \tag{3.41a}$$

$$\boldsymbol{p}_{R_i} = \left[l_a\ 0\ 0\right]^\top \tag{3.41b}$$

$${}^B\mathcal{I}_t = \mathcal{I}_t + m_t||\boldsymbol{p}_t||^2\mathbb{1}_3 - \boldsymbol{p}_t{\boldsymbol{p}_t}^\top \tag{3.41c}$$

$${}^B\mathcal{I}_{R_i} = \mathcal{I}_R + m_{R_i}||\boldsymbol{p}_{R_i}||^2\mathbb{1}_3 - \boldsymbol{p}_{R_i}{\boldsymbol{p}_{R_i}}^\top \tag{3.41d}$$

$$\boldsymbol{R}_{BR_i} = \boldsymbol{R}_z(\gamma_i)\boldsymbol{R}_z(\theta_i)\boldsymbol{R}_x(\beta_i) \tag{3.41e}$$

$$\mathcal{I} = \mathcal{I}_c + \sum_{i=1}^{n}(\boldsymbol{R}_{BR_i}({}^B\mathcal{I}_t + {}^B\mathcal{I}_{R_i}){\boldsymbol{R}_{BR_i}}^\top). \tag{3.42}$$

Table 3.2 System inertia parameters and example values to be used for morphology comparison

property	unit	value
$m_{c,\text{const}}$	[kg]	1.42
m_a	[kg]	0.20
$m_{t,\text{norm}}$	[kgm^{-}1]	0.10
m_{r_i}	[kg]	0.20
L	[m]	0.30

property	unit	value
r_c	[m]	0.05
h_c	[m]	0.02
r_1	[m]	0.0075
r_2	[m]	0.008
r_{r_i}	[m]	0.03
h_{r_i}	[m]	0.065

Table 3.2 contains example inertia values is inspired by the system components sized for the prototypes described in Chap. 5, and is used to generate results of the following morphology comparison in Fig. 3.9 and Table 3.4.

3.2.3 *Morphology Optimization*

In this section, we explore the effect of morphology choice by comparing the capabilities of several compelling models. We develop an optimization problem that considers the desired system capabilities for aerial interaction tasks, as described in Chap. 2, and evaluate the optimization results against some state-of-the-art fully actuated system models. The contents of this section are based on work published in [4].

In defining the problem of aerial manipulation, we proposed that the tilt-rotor omnidirectional MAV offers compelling advantages toward the goal of versatility in force exertion and hover efficiency. The extent to which the system achieves these goals is dependent on rotor group position and tilt-able axes, and as a result, morphology design is important to ensure that the resulting platform meets performance requirements.

We aim to achieve the following design goals:

- Fully actuated system in any hover orientation.
- High force and torque capabilities in all directions.
- High-efficiency hover in at least one orientation.

To evaluate the dynamic capabilities of force- and pose-omnidirectionality, as well as hover efficiency, we define a series of evaluation metrics:

- Force and torque actuation volumes.
- Hover efficiency.
- Overall inertia.

Force and Torque Actuation Volumes

We evaluate the body wrench actuation capabilities of the system using force and torque envelopes described in Sect. 3.1. Torque volumes are computed in the presence of a hover force along e_z^B. Maximum, minimum, mean surface radius, and volume for the envelopes are calculated ($\{f_{\text{max}}, f_{\text{min}}, f_{\text{mean}}, f_{\text{vol}}\}$ and $\{\tau_{\text{max}}, \tau_{\text{min}}, \tau_{\text{mean}}, \tau_{\text{vol}}\}$), and are used to compare dynamic capabilities between morphologies.

Hover Efficiency

The hover efficiency is evaluated by computing the force efficiency index, as defined in Eq. (2.12), as originally defined in [5]. The index represents the ratio of the desired body force magnitude to the sum of individual rotor group thrust magnitudes. When $\eta_f = 1$, no internal forces are present, and all acting forces are aligned with the desired force vector $\boldsymbol{f}^*_{\text{act}}$. While these internal forces should be reduced for efficient flight, they also allow for instantaneous disturbance rejection, since thrust vectoring can be achieved by changing only rotor speeds.

A *torque efficiency index* can be computed similarly in Eq. (3.43) to evaluate the efficiency of maximum torque commands, considering the body moment due to propeller forces acting at distance l_a from O_B. Torque contributions due to rotor drag are assumed to be negligible, since they are a significantly smaller multiple of the thrust force.

$$\eta_\tau = \frac{||\,{}^B\boldsymbol{\tau}^*_{\text{act}}||}{l_a \sum_{i=1}^{n_r} {}^B\boldsymbol{f}_{R_i}} \tag{3.43}$$

Overall Inertia

The dynamic motion ability is affected by its system inertia. We use the approximate inertial formulation in Eq. (3.42) to compare these properties for optimized and state-of-the-art systems.

Reducing inertia increases agility, and can be achieved by reducing the total mass m, and bringing the mass closer the base origin O_B. Dynamic tilt-rotor inertial effects are neglected for this analysis.

Optimization

We present a strategy for parametric optimization of a generalized tilt-rotor model as shown in Fig. 3.7, subject to a cost function of force and torque exertion, agility and efficiency metrics. We define fixed angles $\boldsymbol{\theta}, \boldsymbol{\beta} \in \mathbb{R}^n$ as $\boldsymbol{e}_z^B$ and $\boldsymbol{e}_y^B$ axis angular deviations from a standard multicopter morphology with arms evenly distributed in

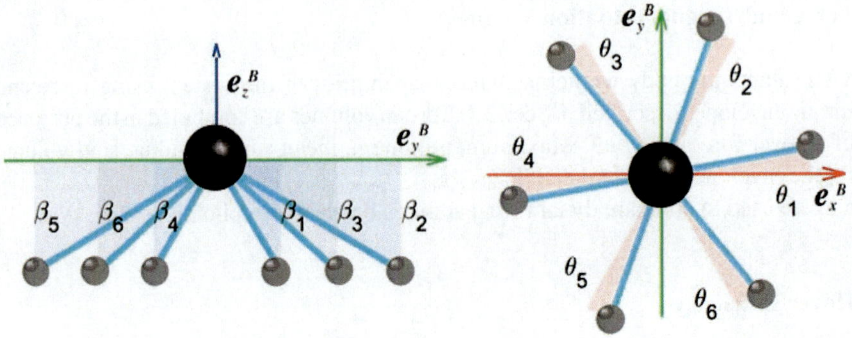

Fig. 3.7 Definitions of angles θ and β for morphology optimization of a tilt-rotor system

the $\boldsymbol{e}_z^B$-plane. Rotor groups can tilt actively about $\boldsymbol{e}_x^{R_i}$, and are controlled independently. For the purposes of this paper, we only optimize over $\boldsymbol{\theta}$ and $\boldsymbol{\beta}$. Although the optimization tool allows for additional optimization over the number of arms n_r and arm length l_a, we set these parameters in advance to 6 and 0.3m respectively.

We consider two cost functions to evaluate an omnidirectional tilt-rotor system: one that prefers unidirectional flight efficiency while requiring omnidirectional flight, and another that maximizes omnidirectional force and torque capabilities. The first cost function is defined as a maximization of the force envelope in one direction (here, unit vector $\boldsymbol{e}_z^B$), while requiring a minimum force in all directions, expressed as

Cost Function 1:

$$\begin{aligned} &\min_{\boldsymbol{\theta},\boldsymbol{\beta}}(-f_{\max}\boldsymbol{e}_z^B) \\ \text{s.t.} \\ &f_{\min} > |mg| \\ &-\frac{\pi}{2} < \theta_i < \frac{\pi}{2} \\ &-\frac{\pi}{2} < \beta_i < \frac{\pi}{2} \\ &\omega_{\min} < \omega_i < \omega_{\max}, \quad \forall i \in \{1 \ldots n\} \end{aligned} \tag{3.44}$$

The second cost function seeks to maximize omnidirectional capability by maximizing the minimum force and torque envelope values in all body directions, i.e. for all surface point directions on a unit sphere. This result will guarantee omnidirectional hover if the system's parameters provide a sufficient thrust-to-weight ratio.

Cost Function 2:

$$
\begin{aligned}
&\min_{\boldsymbol{\theta},\boldsymbol{\beta}}(-f_{\min}, -\tau_{\min}) \\
&\text{s.t.} \\
&f_{\min} > mg \\
&-\frac{\pi}{2} < \theta_i < \frac{\pi}{2} \\
&-\frac{\pi}{2} < \beta_i < \frac{\pi}{2} \\
&\omega_{\min} < \omega_i < \omega_{\max}, \quad \forall i \in \{1 \dots n\}
\end{aligned}
\tag{3.45}
$$

The resulting optimized morphologies and metrics are presented in Table 3.3. The first optimization function results in a tilt-rotor that takes a standard hexacopter morphology. Force in a single direction is maximized along $\boldsymbol{e}_z^B$, and system properties are sufficient to ensure hover in any orientation. The second optimization has multiple solutions of the same result metrics, placing rotor groups O_{R_i} at the vertices of an arbitrarily oriented octahedron. We list the $\boldsymbol{\beta}$ values when $\boldsymbol{\theta}$ are 0. Results show that both systems can sustain omnidirectional hover with additional force capability, and have omnidirectional force and torque envelopes, as seen by the $f_{\min}, \tau_{\min}$.

We further evaluate the evolution of performance metrics between solutions, with $\boldsymbol{\theta}$ values fixed at 0, and $\boldsymbol{\beta} = \beta \cdot [1, -1, 1, -1, 1, -1]$ for changing β. Results are shown in Fig. 3.8, where we see a clear maximization of the minimum reachable force in the left plot. Considering the goal of versatility for efficient flight with omnidirectional capabilities, the $\beta = 0$ solution presents high reachable forces with a maximum efficiency index, and sufficiently large minimum reachable forces and torques for agile flight in 6 DOF.

Table 3.3 Optimized morphologies and simplified metrics

Property	Cost function 1, Eq. (3.44)	Cost function 2, Eq. (3.45)
$\boldsymbol{\beta}$ [°]	[0, 0, 0, 0, 0, 0]	$35.26 \cdot [1, -1, 1 - 1, 1 - 1]$
$\boldsymbol{\theta}$ [°]	[0, 0, 0, 0, 0, 0]	[0, 0, 0, 0, 0, 0]
mass [kg]	4.0	4.0
inertia* [kgm^2]	{0.059, 0.059, 0.117}	{0.078, 0.078, 0.080}
$f_{\{\min,\max\}}$ [N]	{61.3, 122.7}	{81.8, 100.2}
f_{vol} [N^3]	$2.1 \cdot 10^6$	$2.6 \cdot 10^6$
$\tau_{\{\min,\max\}}$ [Nm]	{20.6, 57.3}	{20.7, 35.0}
τ_{vol} [Nm^3]	$1.0 \cdot 10^5$	$7.1 \cdot 10^4$
$\eta_{f,\{\min,\max\}}$ at hover	{0.75, 1.0}	{0.82, 1.0}

*Primary components of inertia are presented, products of inertia are assumed negligible

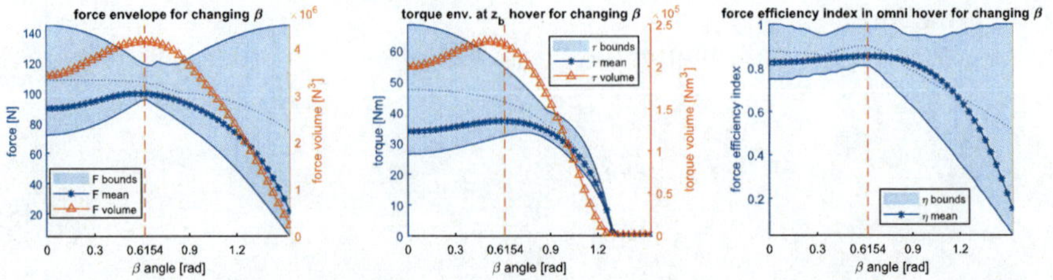

Fig. 3.8 Variation of force and torque envelope metrics (max and min bounds, mean and volume) with changing β, the most efficient hover solution is at $\beta = 0$ and the maximized force/torque envelope is at $\beta = 0.6154$ rad (35.26 deg)

Design Comparison

We compare capabilities of the two optimized tilt-rotor results to other state-of-the-art fully actuated aerial robots.

Despite their advantages in versatility, tilt-rotor systems also come with drawbacks compared to fully actuated fixed rotor systems, so we include both system types in our comparison. Additional actuation mass and complexity due to tilting rotor groups, limited rotation due to possible arm cable wind-up, and the presence of kinematic singularities are not encountered in a fixed rotor system.

Morphologies selected for comparison are as follows:

(a) The flat tilt-rotor result of cost function 1, Eq. (3.44).
(b) The tetrahedrally spaced tilt-rotor result of cost function 2, Eq. (3.45).
(c) State-of-the-art fixed rotor *tilt-hex* realized for aerial interaction applications [6].
(d) State-of-the-art fixed rotor omnidirectional platforms derived and realized by [7, 8].

Morphological model representations for these systems are shown in the far right column of Fig. 3.9.

For fairness of comparison, the lack of tilting motors and related components for fixed rotor morphologies are reflected in reduced mass and rotational inertia, as seen in Table 3.4. The total system mass is estimated to decrease by 30%, but as we assume tilting motors are located close to the platform center, the decrease in rotational inertia is minor. Further differences in mass of fixed and tilt-rotor systems (e.g. battery size) are not estimated here, due to a strong dependency on the detailed design and available technology. Instead, we focus on metrics of platforms with equivalent rotors in different morphological combinations. The remaining affected metrics are the reachable torques which are subject to the gravitational force mg. All systems are modelled with single rotor groups that spin in alternating directions for adjacent groups, and are only invertible for model (d).

For all platforms compared in Fig. 3.9, we show result metrics in Table 3.4 relative to the first tilt-rotor platform. Force and torque envelopes are coloured with the efficiency index of the maximum achievable values in each direction. The third

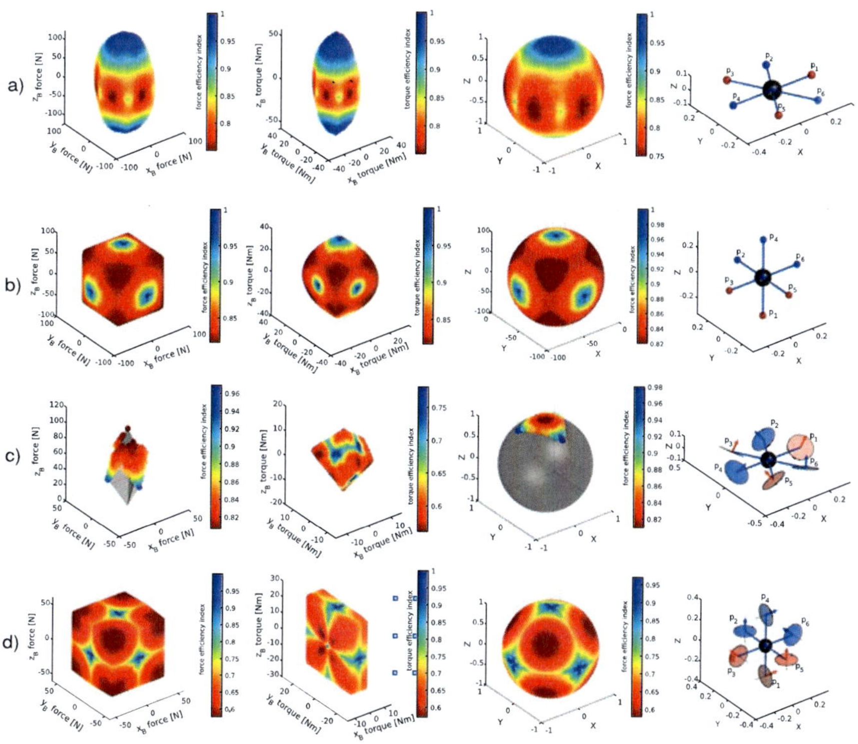

Fig. 3.9 Columns from left to right: reachable force envelope, reachable torque envelope at e_z^B hover (omnidirectional hover directions coloured by force efficiency index), all expressed in $\mathcal{F}_B$, as well as a morphology design illustration are shown for each system. Compared morphologies are (a) a tilt-rotor optimized for unidirectional efficiency, (b) a tilt-rotor that maximizes $f_{\min}$ and $\tau_{\min}$, (c) a fixed rotor design from [6], and (d) a fixed rotor design from [7, 8]. Even numbered propellers (depicted in blue) spin counter-clockwise and odd numbered propellers (red) spin clockwise for positive thrust values. Note that scales vary between plots, which are enlarged to better show the envelope shape. Corresponding numerical metrics are listed in Table 3.4

column shows the force efficiency index for each achievable hover direction, plotted on a unit sphere. As expected, the reachable force and torque envelopes for the two tilt-rotor systems are much larger than their fixed rotor counterparts. Designs (a) and (c) achieve high forces in level hover, while designs (b) and (d) show a more uniform distribution of omnidirectional force. Due to its tilting rotors, design (a) still has better omnidirectional force and torque characteristics than the fixed rotor design (d) which is optimized for omnidirectional force.

The tilt-rotor implementation of a standard hexacopter design promises a versatile and capable morphology solution. For tasks that demand such versatility, the additional weight and complexity of tilt-able rotors can be justified by significantly improved reachable force and torque volumes, and the ability to maintain efficient hover in certain orientations. Independent tilting of each rotor group results in over-

Table 3.4 Comparison of relative omnidirectional system properties corresponding to morphologies in Fig. 3.9

Property*	Tilt-rotor (a)	Tilt-rotor (b)	Fixed rotor (c)	Fixed rotor (d)
Inertia** $\begin{pmatrix} m \\ I_{xx} \\ I_{yy} \\ I_{zz} \end{pmatrix}$	Reference	$\begin{pmatrix} 1.0 \\ 1.32 \\ 1.32 \\ 0.68 \end{pmatrix}$	$\begin{pmatrix} 0.70 \\ 0.99 \\ 0.99 \\ 0.99 \end{pmatrix}$	$\begin{pmatrix} 0.70 \\ 1.31 \\ 1.31 \\ 0.67 \end{pmatrix}$
Force $\begin{pmatrix} min \\ max \\ vol \end{pmatrix}$	Reference	$\begin{pmatrix} 1.33 \\ 0.82 \\ 1.20 \end{pmatrix}$	$\begin{pmatrix} 0 \\ 0.81 \\ 0.02 \end{pmatrix}$	$\begin{pmatrix} 0.67 \\ 0.58 \\ 0.26 \end{pmatrix}$
Torque $\begin{pmatrix} min \\ max \\ vol \end{pmatrix}$	Reference	$\begin{pmatrix} 1.01 \\ 0.61 \\ 0.69 \end{pmatrix}$	$\begin{pmatrix} 0.27 \\ 0.18 \\ 0.01 \end{pmatrix}$	$\begin{pmatrix} 0.30 \\ 0.52 \\ 0.20 \end{pmatrix}$
η_f at hover $\begin{pmatrix} min \\ max \end{pmatrix}$	Reference	$\begin{pmatrix} 1.09 \\ 1.0 \end{pmatrix}$	$\begin{pmatrix} 1.08 \\ 0.98 \end{pmatrix}$	$\begin{pmatrix} 0.77 \\ 1.00 \end{pmatrix}$

* All values are expressed relative to the first tilt-rotor system
**Primary components of inertia are presented, products of inertia are assumed negligible

actuation: 12 inputs to control 6 DOF. The controller can act in the null space of the allocation to assign secondary tasks. We can further justify the tilt-rotor version of a standard hexacopter morphology for perception applications, where the dual unobstructed hemispheres of the $\boldsymbol{e}_z^B$ plane allow for a large field of view.

An example of inertia-normalized acceleration envelopes based on estimated differences is presented in Table 3.5. While various design choices drive the mass difference between fixed and tilt-rotor systems, acceleration envelopes (mass- and inertia-normalized force and torque envelopes) show the advantages of lower actuation weight for fix-tilt systems. These calculations are subject to dramatic change with the choice of actuators, and should be re-evaluated for any specific comparison based on detailed system components.

For the final design, we choose to augment the system with double propeller groups to increase total thrust while maintaining a compact system size. Double rotor groups have the beneficial effect of reducing rotational inertia about the tilting axis, and counter-rotating propellers in each rotor group further reduce unmodelled gyroscopic effects. Upper rotor directions alternate with adjacent arms to reduce unmodelled effects of dual rotor groups related to airflow interaction.

Coordinate systems for the platform are described in Fig. 3.10, with the body and rotor group frames $\mathcal{F}_B$, $\mathcal{F}_{R_i}$, and definitions of the fixed arm spacing angles $\gamma = \frac{\pi}{6} \cdot [1, 3, 5, 7, 9, 11]$ and tilting angles α_i. Individual rotor angular velocities ω_i for $i \in (1, 6)$ represent upper rotors and $i \in (7, 12)$ represent lower rotors (Fig. 3.11).

Table 3.5 Comparison of acceleration envelope metrics corresponding to morphologies in Fig. 3.9

property*	Tilt-rotor (a)	Tilt-rotor (b)	Fixed rotor (c)	Fixed rotor (d)
Inertia** $\begin{pmatrix} m \\ I_{xx} \\ I_{yy} \\ I_{zz} \end{pmatrix}$	Reference	$\begin{pmatrix} 1.0 \\ 1.32 \\ 1.32 \\ 0.68 \end{pmatrix}$	$\begin{pmatrix} 0.70 \\ 0.99 \\ 0.99 \\ 0.99 \end{pmatrix}$	$\begin{pmatrix} 0.70 \\ 1.31 \\ 1.31 \\ 0.67 \end{pmatrix}$
Linear accel $\begin{pmatrix} min \\ max \\ vol \end{pmatrix}$	Reference	$\begin{pmatrix} 1.33 \\ 0.82 \\ 1.20 \end{pmatrix}$	$\begin{pmatrix} 0 \\ 1.15 \\ 0.06 \end{pmatrix}$	$\begin{pmatrix} 0.95 \\ 0.82 \\ 0.75 \end{pmatrix}$
Angular accel $\begin{pmatrix} min \\ max \\ vol \end{pmatrix}$	Reference	$\begin{pmatrix} 0.76 \\ 0.90 \\ 0.59 \end{pmatrix}$	$\begin{pmatrix} 0.20 \\ 0.35 \\ 0.02 \end{pmatrix}$	$\begin{pmatrix} 0.23 \\ 0.79 \\ 0.18 \end{pmatrix}$

*All values are expressed relative to the first tilt-rotor system

**Primary components of inertia are presented, products of inertia are assumed negligible

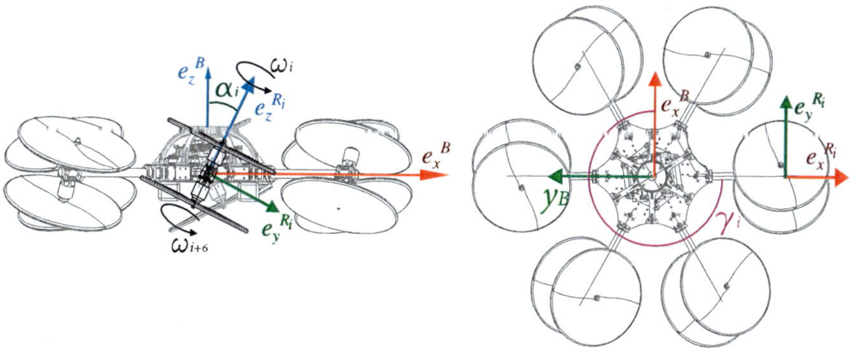

Fig. 3.10 Coordinate frames and variables for the selected tilt-rotor morphology. Principal axes are shown for $\mathcal{F}_B$ and $\mathcal{F}_{R_i}$, as well as rotor group tilt angle α_i, rotor angular velocity ω_i, and arm spacing angle γ_i.

3.3 Aerial Interaction Model

In this section, we define a simple model for interaction with the environment, and develop the momentum-based wrench estimation model that will be used for interaction control. For the sake of general system stability, we assume that the environment is passive, *i.e.* not introducing additional energy into the system. A few additional assumptions define the system's interaction with the environment, framed around our intended aerial interaction tasks.

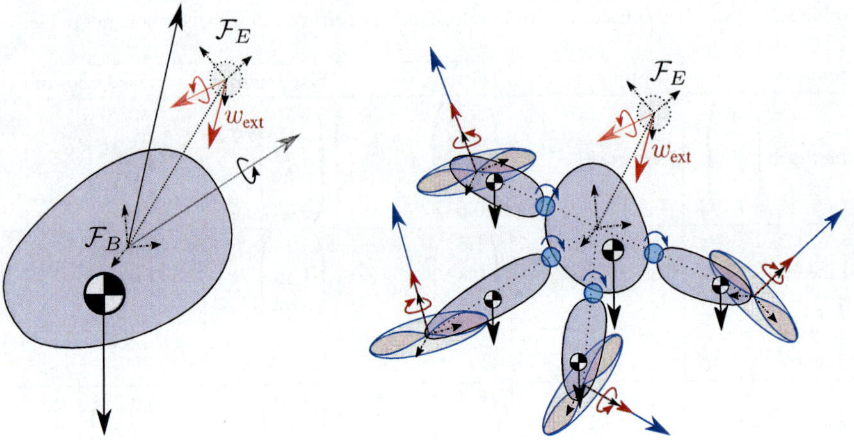

Fig. 3.11 The aerial interaction model as seen for the single rigid body approximation of a tilt-rotor as used for our control methods, and a multi-body model for future extensions to full body interaction control

Assumptions:

- Contact occurs as a wrench applied at a single end-effector point.
- The end-effector has a point-contact with the environment, and negligible surface friction.

In the general case of n_c contact points with the environment where an external wrench $\boldsymbol{w}_{c_i}$ is applied at an arbitrary location c_i on the system, the total external forces described in Eq. (3.3) are expressed as the sum of these external wrenches projected through their corresponding contact Jacobians $\boldsymbol{J}_{c_i}$:

$$\widetilde{\boldsymbol{\tau}}_{\text{ext}} = \sum_{i=1}^{n_c} \boldsymbol{J}_{c_i}^\top \boldsymbol{w}_{c_i}. \tag{3.46}$$

In our assumption that the external wrench occurs at a single defined end-effector location, we simplify Eq. (3.46) to

$$\widetilde{\boldsymbol{\tau}}_{\text{ext}} = \boldsymbol{J}_E^\top \, {}^E\boldsymbol{w}_{\text{ext}}. \tag{3.47}$$

The contact wrench ${}^E\boldsymbol{w}_{\text{ext}} \in \mathbb{R}^6$ expressed in the end-effector frame is projected as generalized forces using the transposed end-effector Jacobian $\boldsymbol{J}_E^\top$.

Contact Constraints

For legged systems such as humanoid and quadruped robots, constraint-consistent dynamics are critical, since the ground contact points are used for propulsion of the system, and the floating base is not directly actuated.

For an aerial flying-base system that is fully actuated by the aerodynamics of rotor-propeller units, the modelling of end-effector contacts can be separated from the primary interest of spatial navigation. Some works have incorporated the dynamic constraints of *e.g.* a wall in order to improve the flying robot's capabilities by pushing against the wall to accelerate more quickly to a desired position [9]. In our assumption that the end-effector has a point-contact with the environment and negligible surface friction during lateral motion, we take a different direction in our model development, prioritizing a representation that is relevant for tasks such as contact inspection and aerial drawing.

Interaction Wrench

Here we provide a preliminary evaluation of the external wrench that will allow us to design a momentum-based external wrench estimator for interaction control in the following chapter. Assuming that model error is negligible, we group disturbance and end-effector contact force into one external wrench in $\mathcal{F}_B$. From Eqs. (3.2) and (3.3), we write the dynamic equations,

$$\mathcal{M}(\boldsymbol{q})\dot{\tilde{\boldsymbol{u}}} + \mathcal{C}(\boldsymbol{q}, \tilde{\boldsymbol{u}})\tilde{\boldsymbol{u}} + \boldsymbol{g}(\boldsymbol{q}) = \tilde{\boldsymbol{\tau}}_{\text{act}} + \tilde{\boldsymbol{\tau}}_{\text{ext}}. \tag{3.48}$$

We consider the know property of mechanical systems [10], that we can express the derivative of $\mathcal{M}$ in terms of $\mathcal{C}$,

$$\dot{\mathcal{M}} = \mathcal{C}(\boldsymbol{q}, \tilde{\boldsymbol{u}}) + \mathcal{C}(\boldsymbol{q}, \tilde{\boldsymbol{u}})^\top, \tag{3.49}$$

and the system's momentum is expressed as

$$\boldsymbol{p} = \mathcal{M}\tilde{\boldsymbol{u}}. \tag{3.50}$$

We consider the physical law of conservation of momentum, and differentiate Eq. (3.50), and substitute in terms from Eqs. (3.48) and (3.49):

$$\dot{\boldsymbol{p}} = \dot{\mathcal{M}}\tilde{\boldsymbol{u}} + \mathcal{M}\dot{\tilde{\boldsymbol{u}}} \tag{3.51}$$

$$= \mathcal{C}(\boldsymbol{q}, \tilde{\boldsymbol{u}})\tilde{\boldsymbol{u}} + \mathcal{C}(\boldsymbol{q}, \tilde{\boldsymbol{u}})^\top\tilde{\boldsymbol{u}} - \mathcal{C}(\boldsymbol{q}, \tilde{\boldsymbol{u}})\tilde{\boldsymbol{u}} - \boldsymbol{g}(\boldsymbol{q}) + \tilde{\boldsymbol{\tau}}_{\text{act}} + \tilde{\boldsymbol{\tau}}_{\text{ext}} \tag{3.52}$$

$$= \mathcal{C}(\boldsymbol{q}, \tilde{\boldsymbol{u}})^\top\tilde{\boldsymbol{u}} - \boldsymbol{g}(\boldsymbol{q}) + \tilde{\boldsymbol{\tau}}_{\text{act}} + \tilde{\boldsymbol{\tau}}_{\text{ext}} \tag{3.53}$$

The final result of Eq. (3.53) will be developed into an estimation method for the external wrench in Sect. 4.2.1.

3.4 Aerial Parallel Manipulator

In this section we present the model of a delta parallel manipulator with 3 translational DOFs attached to an omnidirectional flying base, as shown in Fig. 3.12. Our flying base is modelled as a single rigid body with an attached multi-body manipulator in this book. The ultimate goal of future work will be to extend this formulation to include the multi-body tilt-rotor dynamics.

The kinematic description of a delta manipulator mounted to a single-body omnidirectional floating base is shown in Fig. 3.13. Using inertial and base frames defined in Table 3.1, we describe $\mathcal{F}_D$ as rigidly attached to the flying base, such that it is centered in $\boldsymbol{e}_x^B$ and $\boldsymbol{e}_y^B$, located a point on the negative $\boldsymbol{e}_z^B$ axis (below the system) and rotated π [rad] about the x-axis of $\mathcal{F}_B$. The transformation of $\mathcal{F}_D$ in $\mathcal{F}_B$ is given by a constant translation, ${}^B\boldsymbol{p}_D \in \mathbb{R}^3$, and orientation, $\boldsymbol{R}_{BD} \in SO(3)$.

In the remainder of this section, we present a *kinematic description* of the delta manipulator, then select its dimensions by solving a *geometry optimization* problem for a target end-effector workspace. We follow this with the *manipulator dynamics*, and propose simplifications that will be practical for control implementations.

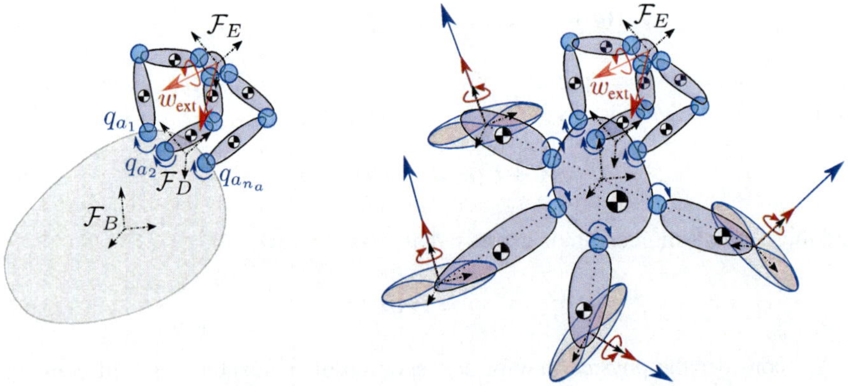

Fig. 3.12 **Left:** the single-body floating base representation used for our system with a multi-body delta manipulator attached. **Right:** The complete multi-body tilt-rotor with parallel manipulator which remains the target of future work

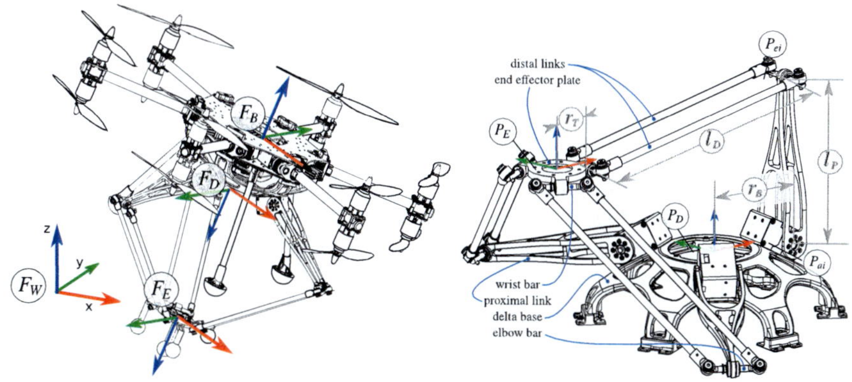

(a) System model including world ($\mathcal{F}_W$), base ($\mathcal{F}_B$), delta base ($\mathcal{F}_D$), and end-effector ($\mathcal{F}_E$) frames.

(b) Parameter descriptions and components of the 3 DoF delta manipulator.

Fig. 3.13 Aerial parallel manipulator model descriptions

3.4.1 Delta Kinematic Description

A visual description of the delta manipulator is provided in Fig. 3.13b. To describe the arm configuration, the actuated joint positions $\boldsymbol{q}_m \in \mathbb{R}^3$ are chosen as generalized coordinates. In the feasible and singularity-free workspace there is a one-to-one correspondence between $\boldsymbol{q}_m$ and the end-effector position with respect to $\mathcal{F}_D$, denoted by ${}^D\boldsymbol{p}_E \in \mathbb{R}^3$.

The kinematic and differential-kinematic functions relating the actuator states (configuration and corresponding velocity) to end-effector states (position and linear velocity) are well known for a 3 DOF delta parallel manipulator [11].

The forward kinematic relation, $h_E^D : \mathbb{R}^3 \rightarrow \mathbb{R}^3$, maps actuated joint positions, $\boldsymbol{q}_m$, to the end-effector position, ${}^D\boldsymbol{p}_E$, such that ${}^D\boldsymbol{p}_E = h_E^D(\boldsymbol{q}_m)$. Geometric constraints require the end psoint of each parallel chain to be at the end-effector point. These constraints are expressed by the loop closure equations $\boldsymbol{\Gamma}_i$ for each parallel chain i, with $i = \{1, 2, 3\}$. For all $i = \{1, 2, 3\}$, it should be that:

$$\boldsymbol{\Gamma}_i := || \, {}^D\boldsymbol{p}_E - {}^D\boldsymbol{p}_{ai} - {}^D\boldsymbol{p}_{ei-ai}(\boldsymbol{q}_m)||^2 - l_D^2 = 0, \tag{3.54}$$

where P_{ai} and P_{ei} represent the point of the joint axis center and elbow frames of the ith parallel chain, respectively. ${}^D\boldsymbol{p}_{ai} \in \mathbb{R}^3$ describes the position of P_{ai}, while ${}^D\boldsymbol{p}_{ei-ai} \in \mathbb{R}^3$ describes the vector between P_{ai} and P_{ei}, both expressed in $\mathcal{F}_D$. Among the solutions of Eq. (3.54), when solving for ${}^D\boldsymbol{p}_E$, we take the one with a greater distance from $\mathcal{F}_D$, since the alternatives would mean passing through an uncontrollable kinematic singularity or violating physical constraints. The closed form solution of Eq. (3.54) is a well-known geometric problem described in [11].

For the forward kinematic derivation, actuator positions are given an inward radial offset equal to the radius of the end-effector plate. This offset effectively reduces the end-effector plate to a point, so that its parallel constraint does not need to be considered. Three spheres are defined with centers at the known positions of the proximal link "elbows", with radius l_D. The 3-sphere intersections represent feasible end-effector positions, and the chosen solution is that with a greater distance from the base, since the alternate would mean passing through an uncontrollable kinematic singularity or violating physical constraints. Inverse kinematics is more straightforward, directly solving the loop closure equations denoted by $\boldsymbol{\Gamma}$. The three loop-closure equations are derived by following each kinematic chain from the base to the end-effector, then virtually closing the loop back to the base. Since the length of the distal link is known, we equate l_D with the norm of the remaining chain and square the result.

For the differential kinematic relations, the end-effector velocity Jacobian is constructed from two matrices obtained by partially differentiating the stacked loop closure equations, $\boldsymbol{\Gamma} = [\boldsymbol{\Gamma}_1\ \boldsymbol{\Gamma}_2\ \boldsymbol{\Gamma}_3]^\top$:

$$\underbrace{\frac{\partial \boldsymbol{\Gamma}}{\partial\, {}^D\boldsymbol{p}_E}}_{\boldsymbol{A}} \delta\, {}^D\boldsymbol{p}_E + \underbrace{\frac{\partial \boldsymbol{\Gamma}}{\partial \boldsymbol{q}_m}}_{\boldsymbol{B}} \delta \boldsymbol{q}_m = 0. \tag{3.55}$$

The resulting Jacobian, $\boldsymbol{J}_D^{D(\boldsymbol{q}_m)} = -\boldsymbol{A}^{-1}\boldsymbol{B} \in \mathbb{R}^{3\times 3}$, maps actuated joint velocities into end-effector velocity, i.e., ${}^D\boldsymbol{v}_E = \boldsymbol{J}_D^D \dot{\boldsymbol{q}}_m$. With three actuators to control 3 DOFs, the manipulator is non-redundant. Two types of kinematic singularities are present [12]: rank loss of $\boldsymbol{A}$ represents an uncontrollable end-effector motion; rank loss of $\boldsymbol{B}$ represents a serial link singularity within at least one kinematic chain, where actuator motion does not lead to end-effector displacement.

Combining the pose and velocity of the flying base, together with Eqs. (3.54) (3.55), we derive the kinematic relations between the position and velocity of the end-effector expressed in $\mathcal{F}_W$, $({}^W\boldsymbol{p}_E, {}^W\boldsymbol{v}_E)$, and the configuration of the aerial manipulator and relative velocity, denoted by $\boldsymbol{q} = (\boldsymbol{q}_b, \boldsymbol{q}_m) \in SE(3) \times \mathbb{R}^3$ and $\dot{\boldsymbol{q}} = [\dot{\boldsymbol{q}}_b^\top\ \dot{\boldsymbol{q}}_m^\top]^\top \in \mathbb{R}^9$, respectively:

$$ {}^W\boldsymbol{p}_E = h_E(\boldsymbol{q}_m) = {}^W\boldsymbol{p}_B + \boldsymbol{R}_{WB}\left({}^B\boldsymbol{p}_D + \boldsymbol{R}_{BD}\, {}^D\boldsymbol{p}_E\right) \tag{3.56}$$

$$ {}^W\boldsymbol{v}_E = \boldsymbol{J}_{PB}\, {}^W\boldsymbol{v}_B + \boldsymbol{J}_{RB}\boldsymbol{\omega}_B + \boldsymbol{J}_D \dot{\boldsymbol{q}}_m = \boldsymbol{J}_E(\boldsymbol{q}_m)\dot{\boldsymbol{q}}_m \tag{3.57}$$

where ${}^D\boldsymbol{p}_E(\boldsymbol{q}_m) = h_E^D(\boldsymbol{q}_m)$ is computed by solving Eq. (3.54), while $\boldsymbol{J}_{PB}$, $\boldsymbol{J}_{RB}$ and $\boldsymbol{J}_D$ are the Jacobians with respect to the components of $\dot{\boldsymbol{q}}_m$, where $\boldsymbol{J}_D = \boldsymbol{R}_{WB}\boldsymbol{R}_{BD}\boldsymbol{J}_D^D$. The Jacobians can be computed differentiating Eq. (3.54) and using Eq. (3.55) to express the linear relations with the components of $\dot{\boldsymbol{q}}_m$.

3.4.2 Delta Geometry Optimization

The drawback of a parallel manipulator is its limited workspace, which we can justify by careful design of the workspace volume, defined by the parameters $\boldsymbol{I}_\Delta = \{l_P, l_D, \bar{r}\}$Since the omnidirectional floating base can achieve any pose, the workspace size must be large enough to compensate for base position error in addition to fast motions which the base dynamics cannot track. We use a genetic optimization algorithm and cost function to select the static manipulator parameters, similar to the procedure described in [13]. Our method differs primarily by the selection of desired workspace geometry. The genetic algorithm approach has been selected in previous works for its robustness and simplicity for a highly nonlinear problem, and we follow suit with our method in this approach.

We describe the optimization problem as follows:

- **Given** a desired volume $\mathbb{V}$.
- **Find** geometric parameters $\boldsymbol{I}_\Delta$ of a delta robot which generate the smallest workspace containing $\mathbb{V}$.

Now it remains to define the workspace geometry, select $\mathbb{V}$, and represent them mathematically for use in an optimization problem.

Delta Workspace

We can describe the manipulator workspace as the intersection of the end-effector workspace for each open chain $i \in \{1 \ldots 3\}$, offset by the end-effector plate radius r_T, such that all open chain endpoints represent the true end-effector location. The radial plate difference is $\bar{r} = r_B - r_T$, assuming that the base plate is larger than the end-effector plate ($\bar{r} > 0$).

We express each chain's workspace in its local frame, *i.e.* rotated to the base actuator frame $\mathcal{F}_{P_{ai}}$ such that the end-effector point is located at ${}^{P_{ai}}\boldsymbol{p}_E = [x; y; z]^\top$. We arrive at the volume definition, h_i, or h_i' when expressed in the local frame:

$$h_i'(x, y, z) = \left[(x - \bar{r})^2 + y^2 + z^2 + l_P^2 - l_D^2\right]^2 - 4l_P^2\left[(x - \bar{r})^2 + z^2\right]^2 \leq 0 \tag{3.58}$$

which represents a spindle torus. We only consider end-effector points that satisfy $\boldsymbol{e}_z^{P_{ai}} > 0$, since negative values imply an impossible inversion of the base and end-effector plates. A diagram of the three intersecting torus geometries is shown in Fig. 3.14.

Therefore, and end-effector point ${}^{D}\boldsymbol{p}_E$ is within the workspace if it satisfies the condition:

$$h_i(\boldsymbol{I}_\Delta, {}^{D}\boldsymbol{p}_E) \leq 0 \quad \text{for } i = 1, 2, 3 \tag{3.59}$$

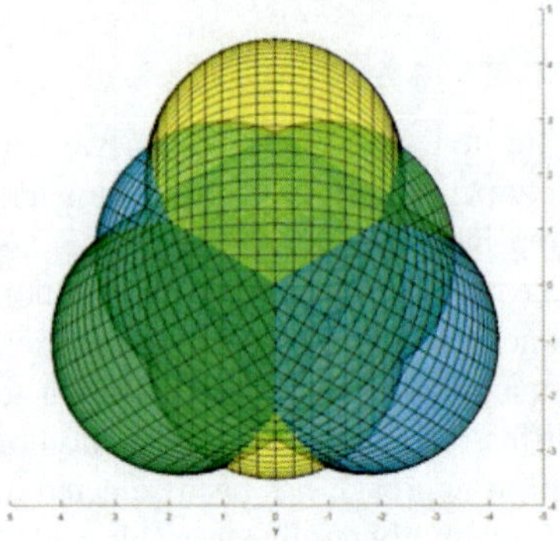

(a) Top view.

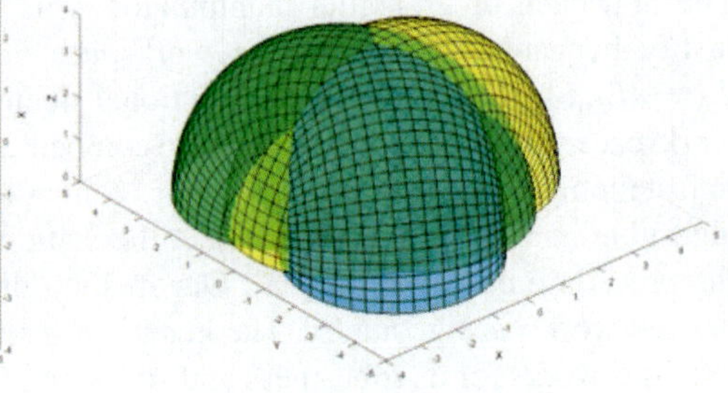

(b) Trimetric view.

Fig. 3.14 Intersecting torus geometries representing the workspace of each delta manipulator chain

Desired Workspace Volume

We choose the desired workspace volume, $\mathbb{V}$, to be a cube with an edge length of 0.2m, which is large enough to accommodate normal flight error at the base while still providing an area for precise end-effector motion. Since the area required for fast motion is highly task-dependent, we choose a size that provides a reasonable space for tracking evaluation without greatly increasing the size of the overall system. Since we also require a compact configuration of the manipulator, we add a retraction point to the workspace of 8.5cm from the nominal base plane, *i.e.* $[0\ 0\ 0.085]^\top$m in $\mathcal{F}_D$. This value is inspired by real design collision constraints. To position the cube in the workspace, we define that it is symmetric along axes $\boldsymbol{e}_x^D$ and $\boldsymbol{e}_y^D$. and positioned such that the nearest cube vertex is at a height c_h above the $\boldsymbol{e}_z^D$ plane. A plot of the desired workspace points inside the torus volume of a single arm chain is shown in Fig. 3.15.

Using the workspace formulation in Eq. (3.59), and considering the workspace to be convex, we state that $\mathbb{V}$ is inside the end-effector workspace if the following holds for each cube vertex and the retraction point, combined as a set of points $\mathbb{P}$ expressed in $\mathcal{F}_D$:

$$h_i(\boldsymbol{I}_\Delta, c_h, \boldsymbol{p}_{\text{ws}}) \leq 0 \quad \text{for } i = 1, 2, 3 \text{ and } \boldsymbol{p}_{\text{ws}} \in \mathbb{P}. \tag{3.60}$$

Genetic Optimization Problem

We formulate the optimization by defining a *fitness function* that will be used by the genetic algorithm. The goal of the fitness function, which should minimize the size of the workspace around the desired volume, thereby reducing the overall manipulator geometry.

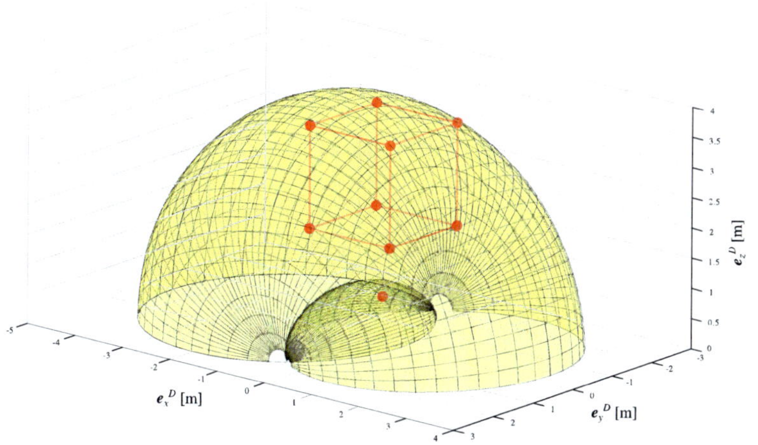

Fig. 3.15 Desired workspace of a cube and retraction point shown in the workspace solution of one delta arm chain

As a metric, we use the *power of point* of a given end-effector point, based on the idea that the value $|h_i(\boldsymbol{I}_\Delta, c_h, \boldsymbol{p}_{\text{ws}})|$ is similar to a distance measurement to the surface, which will be closer to zero for feasible points that contain the solution. The power of point function $f_P(\cdot)$ for point $\boldsymbol{p}$ is defined as

$$f_P(\boldsymbol{p}) = |h_1(\boldsymbol{I}_\Delta, c_h, \boldsymbol{p}_{\text{ws}})| + |h_2(\boldsymbol{I}_\Delta, c_h, \boldsymbol{p}_{\text{ws}})| + |h_3(\boldsymbol{I}_\Delta, c_h, \boldsymbol{p}_{\text{ws}})| \tag{3.61}$$

The geometric synthesis problem can now be formulated as

$$\begin{aligned} &\min_{\boldsymbol{I}_\Delta, c_h} f_P(\boldsymbol{I}_\Delta, c_h, \boldsymbol{p}_{\text{ws}}) \\ &\text{s.t.} \;\; h_i(\boldsymbol{I}_\Delta, c_h, \boldsymbol{p}_{\text{ws}}) \; \forall i \in 1, 2, 3; \; \boldsymbol{p}_{\text{ws}} \in \mathbb{P}. \end{aligned} \tag{3.62}$$

We enforce the additional constraint of $l_D < l_P$ to correspond with a larger end-effector workspace per physical manipulator dimensions. Further details of the fitness function implementation follow the methods in [13].

The genetic optimization is solved with the MATLAB Optimization Toolbox, using a population size of 70, and maximum number of generations of 150.

The final geometric parameters of the delta manipulator resulting from a genetic optimization problem are listed in Table 3.6, where the tool and base plate radius are only specified in terms of their difference $\bar{r}$. The final values for r_B and r_T are selected based on space available at the base, and a required tool plate dimension.

Table 3.6 Delta manipulator parameter descriptions, the genetic algorithm optimized values, and chosen plate radius values

Symbol	Definition	Opt value [m]
l_P	Proximal link length	0.156
l_D	Distal link length	0.024
r_B	Plate radius difference	0.046
c_h	$\mathbb{V}$ min height	0.19
r_B	Delta base plate radius	0.084
r_T	End-effector plate radius	0.038

Manipulator Capabilities

With a defined geometry, we can now evaluate the manipulator's capabilities across the workspace using the manipulator Jacobian $\boldsymbol{J}_D$.

Plots in Fig. 3.16 show the condition number of the manipulator Jacobian, $\boldsymbol{J}_D$, and its component matrices $\boldsymbol{A}$ and $\boldsymbol{B}$ as defined in Eq. (3.55). We define a workspace

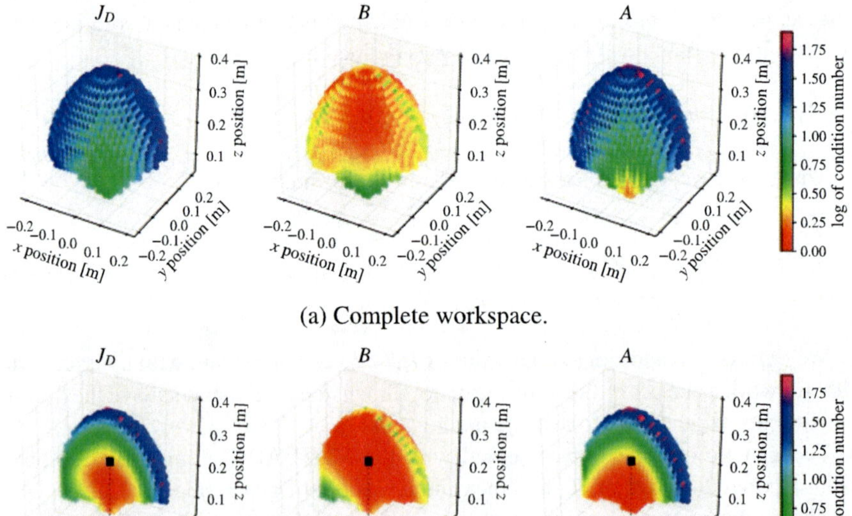

(a) Complete workspace.

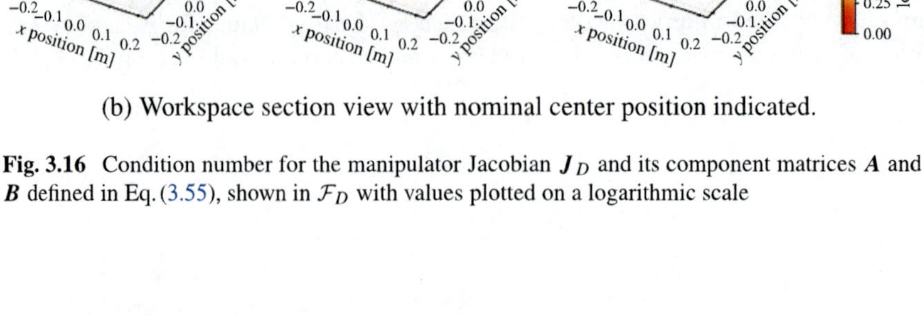

(b) Workspace section view with nominal center position indicated.

Fig. 3.16 Condition number for the manipulator Jacobian $\boldsymbol{J}_D$ and its component matrices $\boldsymbol{A}$ and $\boldsymbol{B}$ defined in Eq. (3.55), shown in $\mathcal{F}_D$ with values plotted on a logarithmic scale center point at $\begin{bmatrix}0 & 0 & 0.25\end{bmatrix}^{\top}$ in $\mathcal{F}_D$ as indicated in Fig. 3.16(b), for its well-conditioned and central location within the reachable workspace.

3.4.3 Aerial Parallel Manipulator Dynamics

Assuming the floating base to be a pure wrench-generating inertial body, and the remaining system to comprise multiple rigid bodies connected by pure rotary joints, the simplified system dynamics are those of a floating base delta manipulator. The nonlinear dynamic model of the combined system can then be derived in closed form, based on recursive algorithms for parallel manipulators.

Inverse Dynamics with Projected Newton-Euler

We consider the equivalent tree structure of the manipulator's parallel chains with a floating end-effector as in [12], and use the PNE method to reflect generalized forces due to dynamics in the platform's base frame. The link connectivity and corresponding open chain structure are shown in Fig. 3.17 where $\mathcal{B}$ refers to a dynamic rigid body link, connected by lines which represent joints.

For the dynamic computation, we group the flying base and delta base links together, and represent the base dynamics directly in $\mathcal{F}_B$. We perform the same iterative PNE process as in Eqs. (3.32) and (3.36), collecting all dynamic terms in the floating base frame. For the end-effector plate, which is connected by a full 6 DOF joint, we transform the dynamic wrench to $\mathcal{F}_B$ with the inverse adjoint transform, $\boldsymbol{Ad}_{\boldsymbol{T}_{BE}^{-1}}$, computed using the manipulator Jacobian $\boldsymbol{J}_D$. Note that to extend the delta

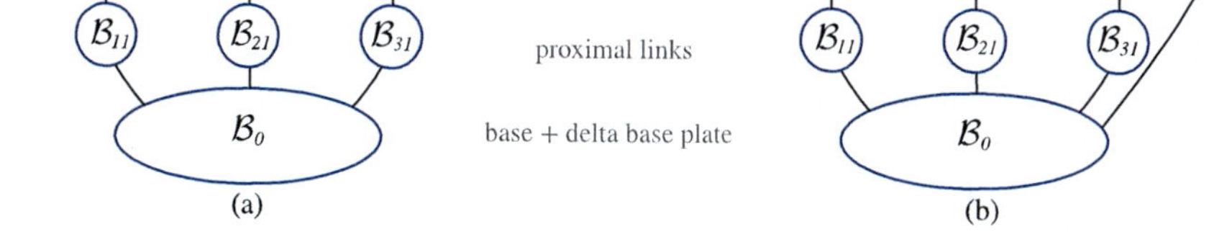

Fig. 3.17 **a** The delta parallel structure, where links $\mathcal{B}$ are connected by lines representing a single-axis rotary joint. **b** The corresponding open chain structure used for dynamic modelling, where a new 6 DOF joint connects the end-effector plate directly to the base

manipulator dynamics to a floating base and to use in the projected Newton-Euler method, the acceleration kinematics must be adapted from those presented in [12] to consider the rotating base frame. The rotated gravitational acceleration is also combined with $\dot{\boldsymbol{\nu}}_B$ as input for dynamic evaluation of the child links to include their gravitational effects. The resulting net dynamic wrench of the multi-body system expressed in $\mathcal{F}_B$ can be expressed as

$$
\begin{aligned}
{}^B\boldsymbol{w}_{B,\text{dyn}} = \boldsymbol{M}(\boldsymbol{q}_b)\dot{\boldsymbol{\nu}}_B + \boldsymbol{C}(\boldsymbol{q}_b, \boldsymbol{\nu}_B)\boldsymbol{\nu}_B + \boldsymbol{g}(\boldsymbol{q}_b) \\
+ \underbrace{\boldsymbol{Ad}^{\top}_{\boldsymbol{T}^{-1}_{BE}} \boldsymbol{w}_{\mathcal{B}_E,\text{dyn}} + \sum_{i=1}^{3} {}^B\boldsymbol{w}_{\mathcal{B}_{i1},\text{dyn}}}_{{}^B\boldsymbol{w}_{\Delta}}
\end{aligned}
\tag{3.63}
$$

where $\boldsymbol{w}_{*,\text{dyn}}$ is the accumulated dynamic wrench of the link $\mathcal{B}*$ and its child links, and ${}^B\boldsymbol{w}_{\Delta}(\ddot{\boldsymbol{q}}, \dot{\boldsymbol{q}}, \boldsymbol{q}) \in \mathbb{R}^6$ represents the coupling wrench between the manipulator and flying base.

Simplified Dynamic Model

The complete model considers the passive manipulator joint dynamics, which results in a highly complex derivation. Observing the system model in Fig. 3.13b, we make the following assumptions for model simplification:

- Elbow and wrist components have a small inertia about their rotational axis and a small change in center of mass position due to rotation. They can therefore be grouped respectively with the corresponding proximal link and the end-effector platform.
- The lightweight distal link bar inertia can be neglected due to a small mass (Thin-walled carbon rods for our final design in Sect. 5.1.5) and inertial effects, with their mass (mostly joint mechanics) divided and grouped to neighbouring bodies.

The resulting simplified model forgoes the need to compute passive joint motion, avoids integration of an internal parallelogram, and greatly reduces the number of computations for the PNE method.

A body link topology of the simplified structure is depicted in Fig. 3.18, where we see the redistribution of link inertia and the resulting open chain topology for dynamic analysis. As with the complete topology, the PNE method is used to generate the same resulting base frame equation as Eq. (3.63), where the summed child link dynamics are now only one layer deep.

The final Lagrangian for of the manipulator dynamics can be derived as described in [12], for use in a whole-body inverse dynamic controller. For the control strategies

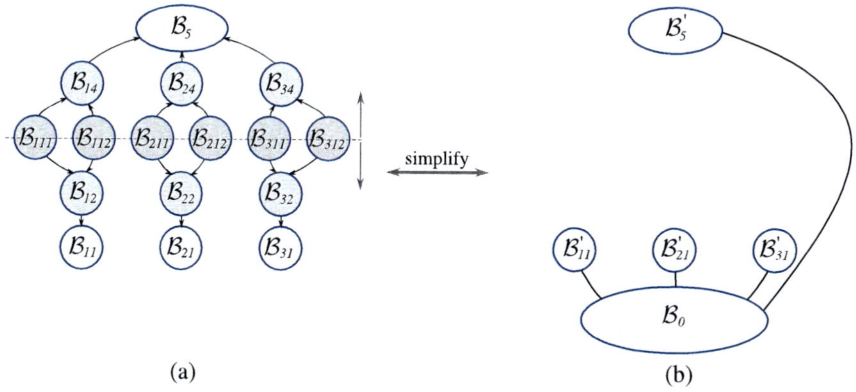

Fig. 3.18 **a** The multi-body dynamic model is simplified by re-distribution of the intermediate links in the indicated directions. The distal links are symmetrically divided and merged with wrist and elbow links. Wrist links are then merged with the end-effector plate and elbow links with corresponding proximal links. **b** The resulting open chain structure for inverse dynamics evaluation, with simplified links labelled as $\mathcal{B}'$

included in this book, it is sufficient to describe the free flight system dynamics in terms of the manipulator reaction wrench:

$$\boldsymbol{M}(\boldsymbol{q}_b)\dot{\boldsymbol{\nu}}_B + \boldsymbol{C}(\boldsymbol{q}_b, \boldsymbol{\nu}_B)\boldsymbol{\nu}_B + \boldsymbol{g}(\boldsymbol{q}_b) + {}^B\boldsymbol{w}_\Delta(\ddot{\boldsymbol{q}}, \dot{\boldsymbol{q}}, \boldsymbol{q}) = {}^B\boldsymbol{w}_{\text{act}}. \tag{3.64}$$

References

1. Stepniewski, W.Z., Keys, C.: Rotary-wing Aerodynamics. Courier Corporation (1984)
2. Brazinskas, M., Prior, S.D., Scanlan, J.P.: An empirical study of overlapping rotor interference for a small unmanned aircraft propulsion system. Aerosp. **3**(4), 32 (2016)
3. Kamel, M., Verling, S., Elkhatib, O., Sprecher, C., Wulkop, P., Taylor, Z., Siegwart, R., Gilitschenski, I.: The voliro omniorientational hexacopter: an agile and maneuverable tiltable-rotor aerial vehicle. IEEE Robot. & Autom. Mag. **25**(4), 34–44 (2018)
4. Allenspach, M., Bodie, K., Brunner, M., Rinsoz, L., Taylor, Z., Kamel, M., Siegwart, R., Nieto, J.: Design and optimal control of a tiltrotor micro-aerial vehicle for efficient omnidirectional flight. Int. J. Robot. Res. **39**(10–11), 1305–1325 (2020)
5. Ryll, M., Bicego, D., Franchi, A.: Modeling and control of FAST-Hex: a fully-actuated by synchronized-tilting hexarotor. In: 2016 IEEE/RSJ International Conference on Intelligent Robots and Systems (IROS), pp. 1689–1694. IEEE (2016)
6. Ryll, M., Muscio, G., Pierri, F., Cataldi, E., Antonelli, G., Caccavale, F., Bicego, D., Franchi, A.: 6D interaction control with aerial robots: the flying end-effector paradigm. Int. J. Robot. Res. 0278364919856694 (2019)
7. Brescianini, D., D'Andrea, R.: Design, modeling and control of an omni-directional aerial vehicle. In: 2016 IEEE International Conference on Robotics and Automation (ICRA), pp. 3261–3266. IEEE (2016)
8. Park, S., Her, J., Kim, J., Lee, D.: Design, modeling and control of omni-directional aerial robot. In: 2016 IEEE/RSJ International Conference on Intelligent Robots and Systems (IROS), pp. 1570–1575. IEEE (2016)

9. Nava, G., Sablé, Q., Tognon, M., Pucci, D., Franchi, A.: Direct force feedback control and online multi-task optimization for aerial manipulators. IEEE Robot. Autom. Lett. **5**(2), 331–338 (2019)
10. Siciliano, B., Sciavicco, L., Villani, L., Oriolo, G.: Robotics: Modelling, Planning and Control. Springer Science & Business Media (2010)
11. Clavel, R.: A fast robot with parallel geometry. In: Proceedings of the International Symposium on Industrial Robots, pp. 91–100 (1988)
12. Briot, S., Khalil, W., et al.: Dynamics of Parallel Robots. Springer, From rigid bodies to flexible elements (2015)
13. Laribi, M., Romdhane, L., Zeghloul, S.: Analysis and dimensional synthesis of the delta robot for a prescribed workspace. Mech. Mach. Theory **42**(7), 859–870 (2007)

Chapter 4
Control

In this chapter, we explore control approaches for aerial manipulation with fully actuated and omnidirectional aerial systems. Based on the models presented in Chap. 3, we focus on the major categories of *free-flight tracking control* with an omnidirectional tilt-rotor flying base in Sect. 4.1, *interaction control* for contact force tracking with a rigidly mounted aerial manipulator in Sect. 4.2, and finally *macro-micro control* of an omnidirectional system with a delta manipulator for fast and precise end-effector tracking in Sect. 4.3.

For all control implementations, we assume that a state estimator provides the current robot state, $\boldsymbol{\xi}$, including position and orientation, linear and angular velocities of the system's base frame, $\mathcal{F}_B$. We can also choose to include rotary joints, $\boldsymbol{q}_t$, and their velocities, $\dot{\boldsymbol{q}}_t$, in the state, which define the current tilt-rotor configuration and motion. With the addition of an actuated manipulator, the coordinates $\boldsymbol{q}_m$ and corresponding velocities $\dot{\boldsymbol{q}}_m$ are also incorporated. We generally express the robot state as the combination of coordinates and generalized velocities, for the coordinate choices presented in Chap. 3.

$$\boldsymbol{\xi} = (\boldsymbol{q}, \widetilde{\boldsymbol{u}}) \tag{4.1}$$

We further assume that a planner provides the controller with a dynamically consistent and feasible trajectory for the state evolution to the acceleration level, which can be sampled at the rate of the controller. For interaction control, the trajectory contains an additional field for desired contact forces. For redundant end-effector control with a macro-micro manipulation system, we assume that the planner provides dynamically consistent trajectories for the floating base and end-effector frames.

The controllers considered in this book are model-based, relying on an accurate dynamic model of the system, and treating any model errors as disturbances to be rejected.

K. Bodie et al., *Omnidirectional Tilt-Rotor Flying Robots for Aerial Physical Interaction*, Springer Tracts in Advanced Robotics 157,
https://doi.org/10.1007/978-3-031-45497-4_4

4.1 Control of an Omnidirectional Flying Base on SE(3)

We consider the free-flight tracking control of a fully actuated and omnidirectional tilt-rotor flying base, which is commanded to follow an arbitrary 6 DOF trajectory in space. Assuming that this trajectory is feasible within the system's limitations, we explore a cascaded control approach, and consider the limitations of the general tilt-rotor system in terms of singularities and configuration-based control authority.

Selecting the control inputs as rotor thrust commands, $\boldsymbol{u}_r \in \mathbb{R}^{n_r}$, and tilting joint commands $\boldsymbol{u}_t \in \mathbb{R}^{n_t}$, we construct the control input vector as

$$\boldsymbol{u} = \begin{bmatrix} \boldsymbol{u}_r \\ \boldsymbol{u}_t \end{bmatrix} \in \mathbb{R}^{n_r+n_t} \tag{4.2}$$

where rotor inputs are considered to be the rotor thrust vector $\boldsymbol{f}_r$ (which is subsequently processed by a low level pulse width modulation (PWM), rotor speed, or rotor force controller), and the tilting joint commands are either tilt joint angles $\boldsymbol{\alpha}$, velocities $\dot{\boldsymbol{\alpha}}$, or torques $\boldsymbol{\tau}_{\alpha}$.

Revisiting the dynamic model for a rigid body flying base in Eq. (3.11), and assuming that disturbance and contact forces are negligible, we phrase the model-based control problem using the generalized acceleration, $\dot{\widetilde{\boldsymbol{u}}}^*$, as our motion objective.

$$\widetilde{\boldsymbol{\tau}}_{\text{act}}(\boldsymbol{u}) = \boldsymbol{\mathcal{M}}(\boldsymbol{q})\dot{\widetilde{\boldsymbol{u}}}^* + \boldsymbol{\mathcal{C}}(\boldsymbol{q}, \widetilde{\boldsymbol{u}})\widetilde{\boldsymbol{u}} + \boldsymbol{g} \tag{4.3}$$

where the generalized actuation forces $\widetilde{\boldsymbol{\tau}}_{\text{act}}$, appear as a nonlinear function of the actuator inputs, $\boldsymbol{u}$, according to the tilt-rotor morphology.

4.1.1 Controller Structures

We consider two types of control structures to generate desired actuator inputs from state feedback information and the dynamic model of the system. An **end-to-end controller structure** takes the desired trajectory, current state estimate and possibly also joint and sensor data as input, and directly computes desired actuator commands for rotor forces and tilting joints. While this type of controller can exploit any over-actuation of the system in the largest sense, and can incorporate constraints on the actuator output directly, the control vector, $\boldsymbol{u}$, appears non-linearly in the equations of motion, and its form is highly dependent on the chosen morphology. As a result, this controller must be specifically tuned to each morphology on which it is implemented.

Instead of an end-to-end solution, if we consider the system model as a single rigid body, in Eq. (4.3), all aerodynamic actuation terms can be grouped as a net actuation wrench acting at the robot base, $\widetilde{\boldsymbol{\tau}}_{\text{act}}(\boldsymbol{u}) = {}^B\boldsymbol{w}_{\text{act}}(\boldsymbol{u})$. If we choose this base actuation wrench, $\bar{\boldsymbol{u}} = {}^B\boldsymbol{w}_{\text{act}}$, as an intermediate control input, we generate a

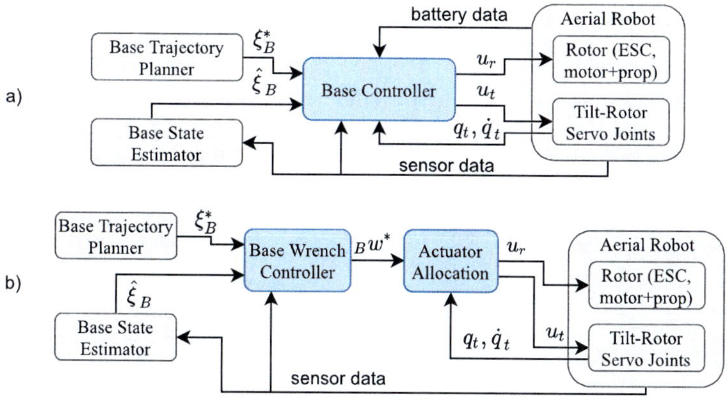

Fig. 4.1 High-level controller structures: **a** end-to-end base controller, and **b** cascaded controller computing an intermediate actuation wrench command for the floating base. Variables ξ_B^* and $\hat{\xi}_B$ indicate desired and estimated state of $\mathcal{F}_B$

cascaded control structure where the input vector appears linearly in the dynamic equations.

$$ {}^B\boldsymbol{w}_{\text{act}} = \bar{\boldsymbol{u}} = \boldsymbol{M}(\boldsymbol{q}_b)\dot{\boldsymbol{\nu}}_B^* + \boldsymbol{C}(\boldsymbol{q}_b, \boldsymbol{\nu}_B)\boldsymbol{\nu}_B + g \tag{4.4} $$

where

$$ \boldsymbol{\nu}_B = \begin{bmatrix} \boldsymbol{v}_B \\ \boldsymbol{\omega}_B \end{bmatrix} \in \mathbb{R}^6 \tag{4.5} $$

A cascaded control structure gives two major system advantages. For one, we now have an equation for which many standard linear controllers can be implemented, which in addition to their uncomplicated implementation, give us a baseline from which to explore and evaluate the behaviour of more complex controllers. Secondly, by interrupting the controller structure with a base wrench computation, the base wrench controller can be platform-agnostic as long as its total actuation capacity is considered by the base wrench controller. A following cascaded block that allocates the desired base wrench to the actuator outputs contains the platform's morphological specificity. Figure 4.1 shows block diagrams of end-to-end and cascaded control structures, with inputs and outputs as they relate to the trajectory planner, state estimation, and the physical robot's actuators.

In the following sections, we explore some implementation options for these controller elements, from generation of the base control wrench and subsequent actuator allocation, to concepts for end-to-end controllers that output actuator commands directly.

4.1.2 Base Wrench Controller

In this section, we consider the first module of the cascaded controller structure mentioned in Sect. 4.1.1, which takes as input the estimated state and reference trajectory, and outputs the computed actuation wrench ${}^{B}\boldsymbol{w}_{\text{act}}$. The result is a state-feedback wrench controller on SE(3) of the single-body system dynamics (modelled in Sect. 3.2), for the base state, $\boldsymbol{\xi}_B$:

$$\boldsymbol{\xi}_B = ({}^{W}\boldsymbol{p}_B, \boldsymbol{R}_{WB}, \boldsymbol{v}_B, \boldsymbol{\omega}_B) \in SE(3) \times \mathbb{R}^6 \tag{4.6}$$

We present in detail the implementation of a PID inverse dynamics controller, and discuss other options for control at a higher level.

PID Control

For underactuated systems such as the standard quadrotor, tracking a position and yaw reference is achieved through cascaded high-level position control and a low-level moment and thrust control loops [1]. Since our omnidirectional tilt-rotor system is fully actuated, we can solve for the wrench control input in one step.

Applying a PID control law, we use the acceleration term as in Eq. (4.4) to represent desired motion for inverse dynamics control:

$$\dot{\boldsymbol{\nu}}^* = \begin{bmatrix} {}^{B}\dot{\boldsymbol{v}}_{\text{PID}} \\ {}^{B}\dot{\boldsymbol{\omega}}_{\text{PID}} \end{bmatrix} = \begin{bmatrix} -\boldsymbol{K}_p \boldsymbol{e}_p - \boldsymbol{K}_i \boldsymbol{e}_{p,i} - \boldsymbol{K}_d \boldsymbol{e}_v + \boldsymbol{R}_{WB}^{\top}\, {}^{W}\dot{\boldsymbol{v}}_B^* \\ -\boldsymbol{K}_{p_R} \boldsymbol{e}_R - \boldsymbol{K}_{i_R} \boldsymbol{e}_{R,i} - \boldsymbol{K}_{d_R} \boldsymbol{e}_\omega + \dot{\boldsymbol{\omega}}_B^* \end{bmatrix}, \tag{4.7}$$

where $\boldsymbol{e}_p, \boldsymbol{e}_v, \boldsymbol{e}_R, \boldsymbol{e}_\omega$ represent the position, velocity, attitude and angular velocity error respectively, and $\boldsymbol{K}_p$, $\boldsymbol{K}_i$, $\boldsymbol{K}_d$, $\boldsymbol{K}_{p_R}$, $\boldsymbol{K}_{i_R}$ and $\boldsymbol{K}_{d_R}$, are corresponding positive definite diagonal gain matrices.

Error terms are formulated in $\mathcal{F}_B$ as follows:

$$\begin{aligned} \boldsymbol{e}_p &= \boldsymbol{R}_{WB}^{\top} \left({}^{W}\boldsymbol{p}_B - {}^{W}\boldsymbol{p}_B^*\right) \\ \boldsymbol{e}_v &= \boldsymbol{R}_{WB}^{\top} \left({}^{W}\boldsymbol{v}_B - {}^{W}\boldsymbol{v}_B^*\right) \\ \boldsymbol{e}_R &= \frac{1}{2}(\boldsymbol{R}_{WB}^{*\top} \boldsymbol{R}_{WB} - \boldsymbol{R}_{WB}^{\top} \boldsymbol{R}_{WB}^*)^{\vee} \\ \boldsymbol{e}_\omega &= \boldsymbol{\omega}_B - \boldsymbol{R}_{WB}^{\top} \boldsymbol{R}_{WB}^* \boldsymbol{\omega}_B^*, \end{aligned} \tag{4.8}$$

with angular error terms defined according to [2]. Here the operator $()^{\vee}$ maps a skew-symmetric matrix to its minimal vector representation in $\mathbb{R}^6$. Integral terms are limited to a value $i_{\max}$ using a saturation function $\text{sat}_{i_{\max}}(\cdot)$, implemented as

$$\boldsymbol{e}_{p,i} = \text{sat}_{i_{p,\max}}\left(\int_t \boldsymbol{e}_p dt\right) \tag{4.9}$$

$$\boldsymbol{e}_{R,i} = \text{sat}_{i_{R,\max}}\left(\int_t \boldsymbol{e}_R dt\right) \tag{4.10}$$

where

$$\text{sat}_{i_{\max}}(y) = \begin{cases} i_{\max} & \text{if} \quad y > i_{\max} \\ y & \text{if} \; -i_{\max} \le y \le i_{\max} \\ -i_{\max} & \text{if} \quad y < -i_{\max} \end{cases} \tag{4.11}$$

Gain matrices act on the error terms and the integrator, where diagonal values correspond to gains on each axis in a chosen frame $\mathcal{F}_{\text{PID}}$,

$$\boldsymbol{K}_p = \boldsymbol{R}_{B\,\text{PID}} \begin{bmatrix} k_{p_x} & 0 & 0 \\ 0 & k_{p_y} & 0 \\ 0 & 0 & k_{p_z} \end{bmatrix} \boldsymbol{R}_{B\,\text{PID}}^{\top} \tag{4.12}$$

Our first option is to express the gain axes in the base frame, i.e. $\boldsymbol{R}_{B\,\text{PID}} = \mathbb{1}_3$, the same frame in which we define the system's force and torque volumes in Sect. 3.1.6. This choice attempts to capture the system's axis-specific control authority as it applies to wrench control. For a system with omnidirectional and time-invariant actuation symmetry, each of the gain matrices could be replaced by a scalar value, since all axes would use equal gains, e.g. $\boldsymbol{K}_p = k_p \mathbb{1}_3$.

In reality, the wrench control authority is complicated by the slower dynamics of tilting rotor groups, and is a nonlinear function of the current tilt-angle configuration. Except in the cases of extremely dynamic flight trajectories, the dominant desired force of the vehicle is due to gravity compensation. The rotors will therefore be oriented to align with the gravity vector, $\boldsymbol{e}_z^W$, as much as possible to maximize flight efficiency. This alignment means that force can be changed with a high bandwidth on the $\boldsymbol{e}_z^W$ axis, regardless of the tilt-angle configuration, but depending on the configuration and system morphology, forces on the orthogonal plane are subject to tilting joint delays. As such, we may benefit from applying linear PID gains in $\mathcal{F}_W$ with more aggressive gain values along $\boldsymbol{e}_z^W$. Alternately, if the robot's morphology gives significantly different actuation capabilities along $\boldsymbol{e}_x^B$ and $\boldsymbol{e}_y^B$, it may be beneficial to use a *heading frame*, $\mathcal{F}_H$, which shares an origin with the base frame $O_H = O_B$, and $\boldsymbol{e}_z^H$ parallel to $\boldsymbol{e}_z^W$.

$$\begin{aligned} {}^W\boldsymbol{e}_z^H &= \boldsymbol{e}_z^W \\ {}^W\boldsymbol{e}_y^H &= \boldsymbol{e}_z^W \times {}^W\boldsymbol{e}_x^B \\ {}^W\boldsymbol{e}_x^H &= {}^W\boldsymbol{e}_y^H \times {}^W\boldsymbol{e}_x^H \end{aligned} \quad \rightarrow \quad \boldsymbol{R}_{B\,\text{PID}} = \boldsymbol{R}_{WB}^{\top} \left[{}^W\boldsymbol{e}_x^H \; {}^W\boldsymbol{e}_y^H \; {}^W\boldsymbol{e}_z^H \right] \tag{4.13}$$

All of the above gain matrices lack incorporation of the system dynamics' inherent dependence on the current tilt-angle configuration. This is a clear limitation of applying constant controller gains, and could be addressed with adaptive configuration-based gains which are computed in the control loop. Since this problem has a higher mathematical complexity, a learning approach could alternately be used, ideally by employing a high-fidelity simulation that includes empirically determined (and perhaps also learned) tilting joint dynamics.

After computing the PID commanded linear and angular acceleration in Eq. (4.7), we substitute these terms into the linear and angular equations of motion derived in Eqs. (3.27) and (3.28) to compute the base wrench control input. The desired actuation forces and torques of the system are expressed as

$$ {}^{B}\boldsymbol{f}^{*}_{\text{act}} = m({}^{B}\dot{\boldsymbol{v}}_{\text{PID}} + {}^{B}g) + \boldsymbol{\omega}_{B} \times m\boldsymbol{v}_{B}. \tag{4.14} $$

$$ {}^{B}\boldsymbol{\tau}^{*}_{\text{act}} = \mathcal{I}_{B}\,{}^{B}\dot{\boldsymbol{\omega}}_{\text{PID}} + \boldsymbol{\omega}_{B} \times \mathcal{I}_{B}\boldsymbol{\omega}_{B} + {}^{B}\boldsymbol{p}_{\circ} \times {}^{B}\boldsymbol{f}_{\text{act}} \tag{4.15} $$

$$ {}^{B}\dot{\boldsymbol{w}}_{\text{act}} = \begin{bmatrix} {}^{B}\boldsymbol{f}^{*}_{\text{act}} \\ {}^{B}\boldsymbol{\tau}^{*}_{\text{act}} \end{bmatrix} \tag{4.16} $$

where ${}^{B}g$ is the gravity acceleration vector along the inertial frame $\boldsymbol{e}_{z}^{W}$ axis, rotated into $\mathcal{F}_{B}$.

In the case where a wrench rate command is required for differential actuator allocation, finite difference is performed to determine the new command at time step k:

$$ {}^{B}\dot{\boldsymbol{w}}_{\text{act}}(t_k) = \frac{{}^{B}\boldsymbol{w}_{\text{act}}(t_k) - {}^{B}\boldsymbol{w}_{\text{act}}(t_{k-1})}{t_k - t_{k-1}} \tag{4.17} $$

4.1.2.1 Optimal Control

Since the previously described PID controller does not incorporate actuation constraints, it is susceptible to degraded performance or failure if the commanded trajectory, excessive control gains or disturbances result in actuator saturation. Formulating the dynamic control problem instead as an optimization problem, we can further define how the wrench command will change over time.

Infinite time formulation: The wrench control problem is formulated as an infinite-time optimal control problem by defining the cost function $\mathcal{J}$ as

$$ \mathcal{J}_{\text{LQRI}} = \text{argmin}_{\bar{\boldsymbol{u}}} \int_{0}^{\infty} ||\boldsymbol{e}(t)||^{2}_{\boldsymbol{Q}} + ||\bar{\boldsymbol{u}}(t)||^{2}_{\boldsymbol{R}} dt \tag{4.18} $$

$$ \text{s.t.}\ \ \dot{\boldsymbol{e}}(t) = f(\boldsymbol{e}(t), \bar{\boldsymbol{u}}(t)) \tag{4.19} $$

where $\boldsymbol{Q} \in \mathbb{R}^{n_e \times n_e}$ and $\boldsymbol{R} \in \mathbb{R}^{n_u \times n_u}$ are weighting matrices for a vector $\boldsymbol{e} \in \mathbb{R}^{n_e}$ containing all state error terms and control input $\bar{\boldsymbol{u}} \in \mathbb{R}^{n_u}$, and $f(\cdot)$ describes the error state evolution based on the system dynamics. Optimal control inputs can be computed e.g. using feedback linearization of the error dynamics with an linear quadratic regulator with integral action (LQRI) gain on the error, as described in detail in [3]. By differentiating the dynamics to the jerk level, we can directly obtain the wrench rate command to be used for differential allocation in the following control block.

Despite its optimal properties, the performance of this type of controller depends on the weights $\boldsymbol{Q}$, $\boldsymbol{R}$ on the state error and inputs, including their proportions relative to each other. While this formulation can make the trade-off between tracking performance and control effort more intuitive compared to PID, compliance with actuator constraints is still not explicitly guaranteed. In this sense, the problem of controller tuning remains key for overall system performance.

Receding horizon formulation: We can instead benefit from our knowledge of the model and future trajectory to implement a receding horizon formulation with MPC, as described in detail in [4]. In this case, the cost function is defined as

$$\mathcal{J}_{\text{MPC}} = \text{argmin}_{\bar{\boldsymbol{u}}} \sum_{k=0}^{N-1} \left(||\boldsymbol{h}_k||^2_{\boldsymbol{Q}} + ||\bar{\boldsymbol{u}}_k||^2_{\boldsymbol{R}} \right) + ||\boldsymbol{h}_N||^2_{\boldsymbol{Q}_N} \tag{4.20}$$

$$\text{s.t.} \quad \boldsymbol{\xi}_k \in \mathbb{S}, \; \bar{\boldsymbol{u}}_k \in \mathbb{U} \tag{4.21}$$

$$\dot{\boldsymbol{\xi}}_k = f(\boldsymbol{\xi}_k, \bar{\boldsymbol{u}}_k) \tag{4.22}$$

$$\boldsymbol{\xi}_0 = \boldsymbol{\xi}(t) \tag{4.23}$$

where $\boldsymbol{h}$ represents a stage cost vector containing state errors $\boldsymbol{e}(\boldsymbol{\xi}, \boldsymbol{\xi}^*)$ and command wrench terms ${}^B\boldsymbol{w}_{\text{act}}$ over the receding horizon of N future time steps, and weighting $\boldsymbol{Q}_N$ is applied to the terminal stage cost at the end of the horizon. The function $f(\cdot)$ now represents the state evolution over time, with $\boldsymbol{\xi}_0$ as the current state (at current time t). The control input vector is chosen as the command wrench rate such that constraints can be applied explicitly at this differential level, and the result can be used directly for differential actuator allocation.

As with the other presented control approaches, the modelling is important for overall task performance, since the predictive accuracy of the controller depends heavily on a good system model. Where system complexities make modelling from first principles extremely difficult, learning-based methods can be used to improve the system model [5].

4.1.3 Instantaneous Actuator Allocation

Following the computation of a desired actuation wrench, ${}^B\boldsymbol{w}^*_{\text{act}}$, the actuator allocation module is responsible for achieving this wrench with commands for the available system actuators. The control allocation method presented here is closely related to work previously presented in [6, 7], going further in depth on aspects of double

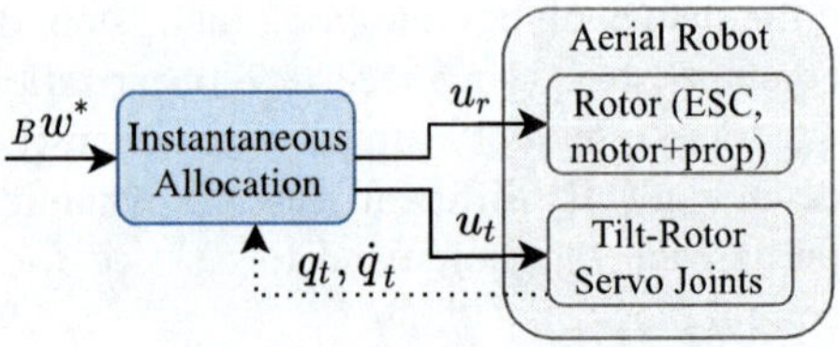

Fig. 4.2 Block diagram of instantaneous actuator allocation

rotor groups and singularity handling. A block diagram of this control component is shown in Fig. 4.2.

Considering a tilt-rotor system with single rotor groups ($n_t = n_r$) and flat morphology ($\beta_i = 0\ \forall i \in \{1...n_\beta\}$), we consider the net actuation wrench from Eq. (3.12) from a control perspective as a nonlinear function of tilt-angle $\boldsymbol{u}_t = \boldsymbol{\alpha}$ and rotor force $\boldsymbol{u}_r = \boldsymbol{f}_r$ inputs.

$$
{}^{B}\boldsymbol{w}_{\text{act}} = \boldsymbol{A}_{\text{act}}(\boldsymbol{u}_t)\boldsymbol{u}_r \tag{4.24}
$$

Writing out $\boldsymbol{A}_{\text{act}}$, we have the following composition per i^{th} column:

$$
\boldsymbol{a}_{\text{act},i} = \begin{bmatrix} \sin(\gamma_i)\sin(\alpha_i) \\ -\cos(\gamma_i)\sin(\alpha_i) \\ \cos(\alpha_i) \\ -s_{r_i}c_d\sin(\gamma_i)\sin(\alpha_i) + l_a\cos(\gamma_i)\cos(\alpha_i) \\ s_{r_i}c_d\cos(\gamma_i)\sin(\alpha_i) - l_a\sin(\gamma_i)\cos(\alpha_i) \\ -l_a\sin(\alpha_i) - s_{r_i}c_d\cos(\alpha_i) \end{bmatrix} \quad \forall i \in \{1...n_t\} \tag{4.25}
$$

All elements of Eq. (4.25) that multiply f_{r_i} are linear combinations of $\sin(\alpha_i)$ and $\cos(\alpha_i)$ due to the associated kinematic rotor group transformation $\boldsymbol{T}_{BR_i}$. We use this property to formulate a variable substitution of intermediate actuator inputs in order to create a linear relationship between control inputs and the net actuation wrench. We subsequently separate each column in Eq. (4.25) into two columns corresponding with the new inputs, $\tilde{\boldsymbol{u}}$, defined as follows:

$$
\bar{\boldsymbol{a}}_{\text{act},i} = \begin{bmatrix} \sin(\gamma_i) & 0 \\ -\cos(\gamma_i) & 0 \\ 0 & \cos(\alpha_i) \\ -s_{r_i}c_d\sin(\gamma_i) & l_a\cos(\gamma_i) \\ s_{r_i}c_d\cos(\gamma_i) & -l_a\sin(\gamma_i) \\ -l_a & -s_{r_i}c_d \end{bmatrix} \qquad \tilde{\boldsymbol{u}} = \begin{bmatrix} \sin(\alpha_1)f_{r_1} \\ \cos(\alpha_1)f_{r_1} \\ \vdots \\ \sin(\alpha_{n_t})f_{r_{n_t}} \\ \cos(\alpha_{n_t})f_{r_{n_t}} \end{bmatrix} \in \mathbb{R}^{2n_t} \tag{4.26}
$$

Collecting the columns $\bar{\boldsymbol{a}}_{\text{act},i}$ in Eq. (4.26) for all n_t rotor groups, we now formulate the time-invariant *static allocation matrix*, $\bar{\boldsymbol{A}}_{\text{act}}$, mapping propeller thrust and torques to the net actuation wrench.

$$^{B}\boldsymbol{w}_{\text{act}} = \bar{\boldsymbol{A}}_{\text{act}}\tilde{\boldsymbol{u}} \qquad \text{where } \bar{\boldsymbol{A}}_{\text{act}} \in \mathbb{R}^{6\times 2n_t} \tag{4.27}$$

The modified input vector $\tilde{\boldsymbol{u}}$ has two values for each tilting rotor group, which are nonlinear combinations of tilt angle and rotor force actuation inputs (α_i, f_{r_i}). These elements of the vector correspond to the lateral and vertical components of each rotor force in its respective frame, $\mathcal{F}_{R_i}$, and we can use trigonometry to reconstruct the true actuator inputs as follows:

$$\begin{aligned} \alpha_i &= \arctan 2\left(\sin(\alpha_i) f_{r_i}, \cos(\alpha_i) f_{r_i}\right) \\ f_{r_i} &= \sqrt{(\sin(\alpha_i) f_{r_i})^2 + (\cos(\alpha_i) f_{r_i})^2}. \end{aligned} \tag{4.28}$$

The formulation in Eq. (4.28) guarantees positive inputs, appropriate for our system with non-invertible propellers. A positive thrust solution is always found, though in certain configurations a small change in force input can lead to large changes in the required tilt-angle orientation. Addressing this condition is the subject of singularity handling, described later in this section.

Actuator control inputs are computed from the base frame command wrench, rearranging Eq. (4.27) using the Moore-Penrose pseudo-inversion (†) of the static allocation matrix, $\bar{\boldsymbol{A}}_{\text{act}}$. We note that we must have $\textbf{rank}\left\{\bar{A}_{\text{act}}\right\} = 6$ for a fully actuated system, which is determined by the tilt-rotor morphology.

$$\tilde{\boldsymbol{u}} = \bar{\boldsymbol{A}}_{\text{act}}^{\dagger}\ {}^{B}\boldsymbol{w}_{\text{act}} \tag{4.29}$$

The pseudo-inverse formulation implicitly minimizes rotor forces in a least-squares sense, generating a tilt-rotor configuration with maximum efficiency. Final actuation inputs $\boldsymbol{u}_r, \boldsymbol{u}_t$ are computed using Eq. (4.28). The resulting α_i values are then saturated according to a tuning variable $\dot{\alpha}_{\text{max}}$. This allocation solution requires minimal computation since the pseudo-inverse of $\bar{\boldsymbol{A}}_{\text{act}}$ is constant and can be performed in advance. As such the solution is suitable for implementation on a low power flight controller.

The present allocation implicitly assumes that tilt angles are achieved instantaneously, which goes against our modelling assumptions. While this formulation is acceptable for slow changes in the actuation wrench, we can adapt the solution using tilt-angle feedback to solve for the current configuration. In this case, $\boldsymbol{u}_t$ are computed as above, but $\boldsymbol{u}_r$ are computed by inverting $\boldsymbol{A}_{\text{act}}$ with the current values $\boldsymbol{\alpha}_{\text{meas}}$.

$$\boldsymbol{u}_r = \boldsymbol{A}_{\text{act}}(\boldsymbol{\alpha}_{\text{meas}})^{\dagger}\ {}^{B}\boldsymbol{w}_{\text{act}} \tag{4.30}$$

We encounter a problem with the formulation in Eq. (4.30), particularly for our chosen system morphology. In certain cases where propellers align $\boldsymbol{A}_{\text{act}}(\boldsymbol{\alpha}_{\text{meas}})$ loses rank, and the pseudo-inversion does not provide a solution, regardless of over-actuation. This equation could be reformulated as an optimization problem with slack variables that allow for an inexact solution. Regardless, this solution requires complex formulation and higher computational power than the solution using Eq. (4.28). In Sect. 4.1.4, we will explore an alternate allocation strategy by taking $\boldsymbol{A}_{\text{act}}$ to the next differential level.

Double Rotor Groups

In the final system design presented in Chap. 5, we choose to implement counter-rotating co-axial rotor groups, such that $n_r = 2n_t$. As modelled in Sect. 3.1.3, two rotor force inputs $f_{r_i}, f_{r_{i+n_t}}$ correspond with one tilt angle α_i for rotor groups $i \in 1 \ldots n_t$. Apart from a change in dimensionality, the matrix $\boldsymbol{A}_{\text{act}}(\boldsymbol{u}_t) \in \mathbb{R}^{6\times 2n_t}$ is still a linear combination of rotor force inputs $\boldsymbol{u}_r \in \mathbb{R}^{2n_t}$.

If we write out the columns of $\boldsymbol{A}_{\text{act}}$ as we did for single rotor groups in Eq. (4.25), a variable substitution can be performed with four values for each tilt arm, with upper rotors $i \in \{1 \ldots n_t\}$ and lower rotors $j \in \{n_t + 1 \ldots 2n_t\}$. We can then separate terms for a static allocation matrix and intermediate inputs for a tilt-rotor system with $n_t = 6$:

$$
{}^{B}\boldsymbol{w}_{\text{act}} = \begin{bmatrix} \bar{\boldsymbol{A}}_{\text{act},i} & \bar{\boldsymbol{A}}_{\text{act},j} \end{bmatrix} \tilde{\boldsymbol{u}} \quad \tilde{\boldsymbol{u}} = \begin{bmatrix} \sin(\alpha_i) f_{r_i} \\ \cos(\alpha_i) f_{r_i} \\ \vdots \\ \sin(\alpha_i) f_{r_j} \\ \cos(\alpha_i) f_{r_j} \\ \vdots \end{bmatrix} \quad \begin{matrix} \forall i \in \{1 \ldots 6\} \\ \forall j \in \{7 \ldots 12\} \end{matrix} \tag{4.31}
$$

where $\bar{\boldsymbol{A}}_{\text{act},i}$ and $\bar{\boldsymbol{A}}_{\text{act},j}$ are formulated as in Eq. (4.26), but differ in the spin direction of upper and lower rotors.

Here we present a method for extracting real inputs from $\tilde{\boldsymbol{u}}$, and will follow with an explanation for the error in this method. To extract α_i values from the result vector $\tilde{\boldsymbol{u}}$, each pair of rows is first summed. Then, we again use the trigonometric identities $\theta = \arctan 2(\sin(\theta), \cos(\theta))$ for tilt angle extraction and $\sin^2(\theta) + \cos^2(\theta) = 1$ for rotor force extraction.

$$
\begin{aligned}
f_{r_i} &= \sqrt{(f_{r_i} \sin(\alpha_i))^2 + (f_{r_i} \cos(\alpha_i))^2} \\
f_{r_{i+6}} &= \sqrt{(f_{r_{i+6}} \sin(\alpha_i))^2 + (f_{r_{i+6}} \cos(\alpha_i))^2} \\
\alpha_i &= \arctan 2((f_{r_i} + f_{r_{i+6}}) \sin(\alpha_i), (f_{r_i} + f_{r_{i+6}}) \cos(\alpha_i))
\end{aligned} \tag{4.32}
$$

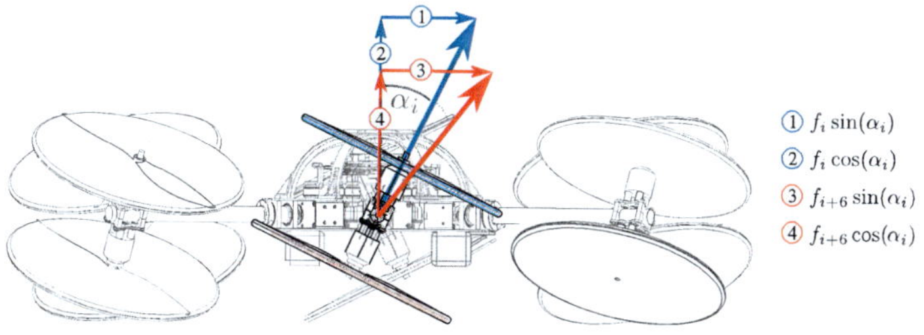

Fig. 4.3 A visual depiction of the over-defined solution for double rotor groups as formulated in Eq. (4.31)

While the solution presented in Eq. (4.32) is technically correct, the formulation of $\tilde{\boldsymbol{u}}$ does not enforce that α_i is the same for upper and lower rotor groups. As a result, the solution of Eq. (4.29) generates 4 equations for 3 unknowns, over-defining α_i. The physical interpretation of this is shown in Fig. 4.3, where each rotor is assigned a different value of α_i.

While it initially appears clever to take advantage of additional over-actuation in a simplified formulation, exposing the differential torque of counter-rotating actuators, this implementation is not sufficient. We also note that the differential torque generated by co-axial rotors is insignificant compared to the net torque resulting from slightly tiling a rotor group. Based on these conclusions, when implementing direct allocation on our double rotor system, we prefer to model the double rotor group as a single, pure thrust rotor group, with negligible drag torque.

Rank Reduction Singularity Handling

We present an approach to handle instantaneous rank reduction singularities discussed in Sect. 3.1.7 for the previously described instantaneous actuator allocation.

The instantaneous singularities are handled using a heuristic approach, identifying the case where the desired force vector ${}^B\boldsymbol{f}^*_{\text{act}}$ comes within an angle φ_t of the $\boldsymbol{e}^B_z$ axis or the $\boldsymbol{e}^B_x$-$\boldsymbol{e}^B_y$ plane. A multiplier k_t then quadratically scales an additional *tilt bias* variable up to a maximum value of c_t as the misalignment angle ϕ goes to zero. A visual depiction of the threshold region defined by φ_t and the changing value c_t is shown in Fig. 4.4. Bias tilt directions b_i alternate for neighbouring tilt-arms. This forced 'toe-in' of neighbouring propeller groups generates constant internal forces, avoiding the singularity by reducing flight efficiency. Writing the change between two subsequent tilt angle commands as $\delta\alpha_i = \alpha_i - \alpha_{i,\text{prev}}$ separated by time step δt, the modified tilt angle $\tilde{\alpha}_i$ can be expressed as

$$\tilde{\alpha}_i = \alpha_{i,\text{prev}} + \text{sign}(\delta\tilde{\alpha}_i)\min(|\delta\tilde{\alpha}_i|, \dot{\alpha}_{\max}\delta t) \qquad \delta\tilde{\alpha}_i = \delta\alpha_i + k_t b_i c_t \tag{4.33}$$

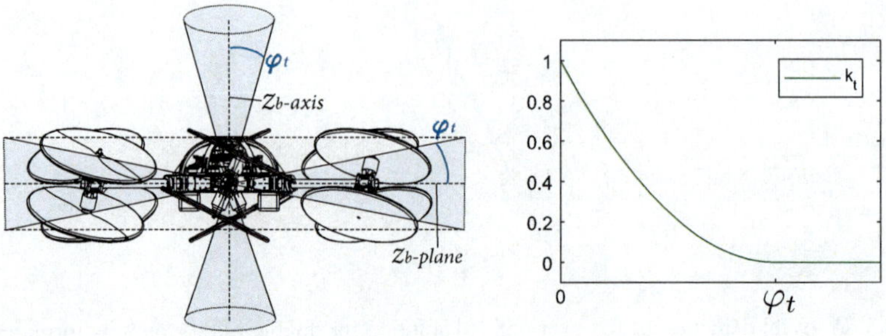

Fig. 4.4 Visual depiction singularity handling for instantaneous rank reduction of $\boldsymbol{A}_{\text{act}}$ with associated multiplier k_t

$$k_t = \begin{cases} 0 & \text{if } \phi \geq \varphi_t \\ (1 - \frac{\phi}{\varphi_t})^2 & \text{if } \phi < \varphi_t \end{cases} \qquad \text{where } b_i = (-1)^i \ \forall \ i \in \{1...6\} \tag{4.34}$$

The resulting condition number envelope for the modified allocation matrix, $\kappa(\boldsymbol{A}_{\text{act}}(\tilde{\alpha}))$, is significantly reduced in the singularity cases, as seen when comparing Fig. 4.5a and Fig. 4.5b, where the scale of (b) is magnified for a better geometric image of the resulting condition number envelope. From the condition number envelopes we see that in the neighbourhood of singular configurations, we can expect deteriorated performance of the system due to low control authority on a certain axis of the control wrench.

Kinematic Singularity Handling

We also present a method for handling kinematic singularities, which are defined in Sect. 3.1.7. An approach commonly used in robot manipulators can be invoked to handle kinematic singularities uses the damped pseudo-inverse of the allocation matrix [8]. This approach, however, requires the outputs in question to be expressed linearly in the allocation, which is not the case for our system. Taking inspiration from this approach, a heuristic solution that approximates a damping effect is derived here.

When the angle between the desired force vector comes within the damping threshold angle φ_d of an arm axis, the tilt-angle speeds associated with that axis are quadratically scaled to zero by multiplier $k_{\alpha,i}$. As the alignment angle η_i approaches a zero threshold, φ_0, the tilt angle stops any motion. A visual interpretation of the alignment geometry is shown in Fig. 4.6 also depicting the multiplier curves $k_{\alpha,i}$ and $(1 - k_{\alpha,i})$. The change in α_i between time steps can then be expressed as follows:

$$\delta\alpha_i' = \delta\alpha_i(1 - k_{\alpha,i}) \tag{4.35}$$

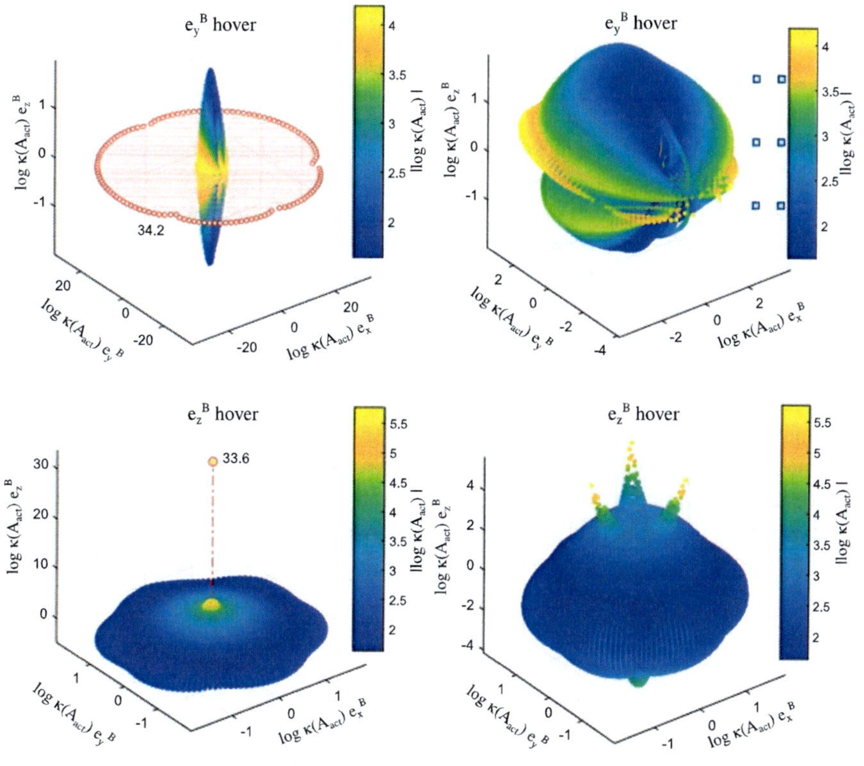

(a) No tilt bias, $\boldsymbol{A}_{\mathrm{act}}(\boldsymbol{\alpha})$.

(b) With tilt bias, $\boldsymbol{A}_{\mathrm{act}}(\tilde{\boldsymbol{\alpha}})$.

Fig. 4.5 Condition number of $\boldsymbol{A}_{\mathrm{act}}(\boldsymbol{\alpha})$ for desired force directions with magnitude mg around the unit sphere combined with a hover force along $\boldsymbol{e}_y^B$, and hover force along $\boldsymbol{e}_z^B$. Values for $\boldsymbol{\alpha}$ correspond with the computed solution **a** without and **b** with applied tilt bias. Scale units are $\log \kappa(\boldsymbol{A}_{\mathrm{act}})$, and axes are magnified **b** to distinguish the resulting geometry

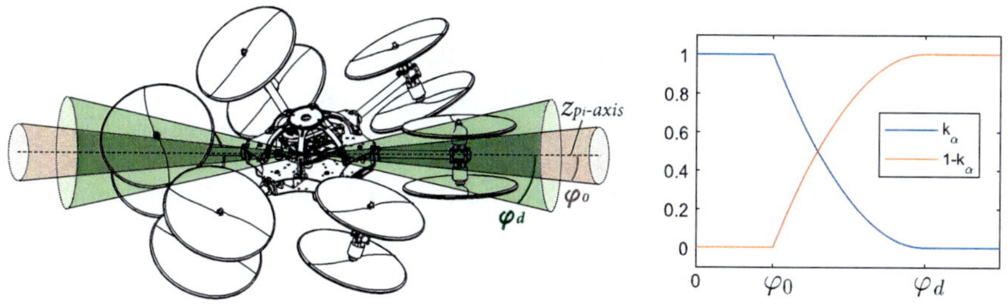

Fig. 4.6 Visual depiction of kinematic singularity handling with associated multiplier k_α

$$k_{\alpha,i} = \begin{cases} 0 & \text{if} \quad \eta_i > \varphi_d \\ (1 - \frac{\eta_i - \varphi_0}{\varphi_d - \varphi_0})^2 & \text{if } \varphi_d > \eta_i > \varphi_0 \\ 1 & \text{if} \quad \eta_i < \varphi_0 \end{cases} \tag{4.36}$$

This approach is justified due to the fact that at small angles of alignment deviation, the force contribution of the associated rotor group is negligible when compared to the remaining thrust contributions, and the error can be treated as a disturbance. When the new angles are computed based on these values, and limited according to $\dot{\alpha}_{\max}$, the rotor forces are then calculated as in Eq. (4.32). A force from the limited propeller group exists only if the thrust axis can be projected as a positive component onto the desired force vector. We note that this heuristic solution is based on the assumption that the command torque is negligible, meaning that for steady-state hover, the system's COM is located approximately at the center of geometry.

Since the contribution of propeller groups in the tilt-angle singularity region is so small, one can take advantage of this situation to address another issue of the physical system, notably the wind-up of cables around an axis. As the tilt velocity is ramped down, it is balanced with an unwinding velocity ω_u, that is calculated based on a maximum unwinding rate and direction to the zero position, resulting in the new tilt angle change.

$$\delta\alpha_i^* = \delta\alpha_i' - \text{sign}(\alpha_i)\, k_{\alpha,i}\omega_u\delta t. \tag{4.37}$$

The resulting tilt angle commands are then computed as before in Eq. (4.33).

4.1.4 Differential Actuator Allocation

The instantaneous allocation approach presented in Sect. 4.1.3, while requiring minimal computational power, suffers from several major drawbacks:

- Tilting velocities are not considered in the allocation.
- Singularities are explicitly handled with a heuristic solution.
- Cable unwinding only occurs in the region of a kinematic singularity.

We address all of these problems simultaneously by differentiating the actuator allocation problem. The content of this section is based on collaborative contributions from [3]. A block diagram of this control component is shown in Fig. 4.7.

Taking the derivative of the actuation wrench, we define the differential allocation matrix from Eq. (3.21) as $\dot{\boldsymbol{A}}_{\text{act}} \in \mathbb{R}^{6\times n_u}$ where $n_u = n_r + n_t$. We reduce this symbol to $\tilde{\boldsymbol{A}}$ to simplify notation and improve clarity in the following equations. The wrench rate command, ${}^B\dot{\boldsymbol{w}}_{\text{act}}$, is the output of a base wrench controller, and the differential inputs, $\dot{\boldsymbol{u}} \in \mathbb{R}^{n_u}$, include rotor force rates and tilting angular velocities.

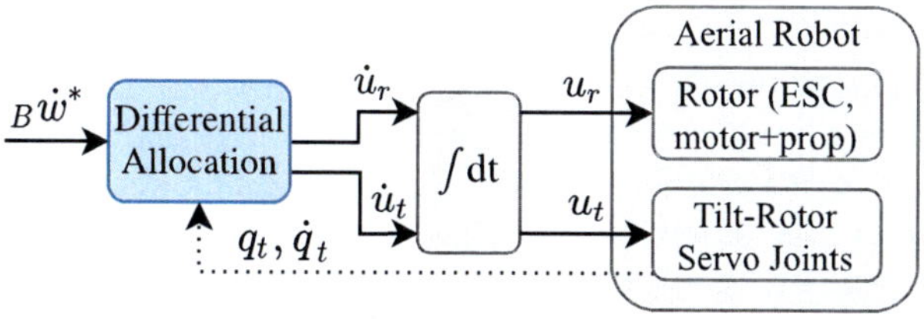

Fig. 4.7 Block diagram of differential actuator allocation

$$ {}^{B}\dot{\boldsymbol{w}}_{\text{act}} = \underbrace{\left[\boldsymbol{A}_{\text{act}}(\boldsymbol{\alpha}) \;\; \frac{\partial \boldsymbol{A}_{\text{act}}(\boldsymbol{\alpha})}{\partial \boldsymbol{\alpha}} \boldsymbol{f}_r\right]}_{\tilde{\boldsymbol{A}}(\boldsymbol{\alpha}, \boldsymbol{f}_r)} \underbrace{\begin{bmatrix} \dot{\boldsymbol{f}}_r \\ \dot{\boldsymbol{\alpha}} \end{bmatrix}}_{\dot{\boldsymbol{u}}} \tag{4.38} $$

Since all actuator inputs now appear explicitly in the actuator input vector, $\dot{\boldsymbol{u}}$, we pose a weighted least squares optimization of an arbitrary desired input vector $\dot{\boldsymbol{u}}^*$ in the differential allocation null space. The optimization problem is framed as

$$ \begin{aligned} &\min_{\dot{\boldsymbol{u}}} ||\boldsymbol{W}(\dot{\boldsymbol{u}} - \dot{\boldsymbol{u}}^*)||_2 \\ &\text{s.t.} \quad {}^{B}\dot{\boldsymbol{w}}_{\text{act}} = \tilde{\boldsymbol{A}}\dot{\boldsymbol{u}} \end{aligned} \tag{4.39} $$

where $\boldsymbol{W} \in \mathbb{R}^{n_u \times n_u}$ is a positive diagonal weighting matrix. The analytical solution to this problem can be expressed as follows:

$$ \dot{\boldsymbol{u}} = \dot{\boldsymbol{u}}^* + \boldsymbol{W}\tilde{\boldsymbol{A}}^\top(\tilde{\boldsymbol{A}}\boldsymbol{W}\tilde{\boldsymbol{A}}^\top)^{-1}({}^{B}\dot{\boldsymbol{w}}_{\text{act}} - \tilde{\boldsymbol{A}}\dot{\boldsymbol{u}}^*). \tag{4.40} $$

We formulate the desired input vector by placing a gain on the error between current values for rotor forces, $\boldsymbol{f}_{r,c}$, and tilt angles $\boldsymbol{\alpha}_c$, against desired values $\boldsymbol{f}_r^*$ and $\boldsymbol{\alpha}^*$ as

$$ \dot{\boldsymbol{f}}_r^* = \text{sign}\left(\boldsymbol{f}_r^* - \boldsymbol{f}_{r,c}\right) c_f v_\omega \tag{4.41} $$

$$ \dot{\boldsymbol{\alpha}}^* = \text{sign}\left(\boldsymbol{\alpha}^* - \boldsymbol{\alpha}_c\right) v_\alpha \tag{4.42} $$

$$ \dot{\boldsymbol{u}}^* = \begin{bmatrix} \dot{\boldsymbol{f}}_r^* \\ \dot{\boldsymbol{\alpha}}^* \end{bmatrix} \tag{4.43} $$

where v_ω and v_α are positive scalar tuning parameters representing a velocity of attraction towards the desired solution. The choice of desired values $\boldsymbol{f}_r^*$, $\boldsymbol{\alpha}^*$ is based on the competing goals of high efficiency flight, and unwinding of tilt-arm cables. The simple solution based on the inverse static allocation in Eq. (4.27) and subsequent actuator command extraction provides both of these elements. For tilt angles, $\boldsymbol{\alpha} =$

0 corresponds with a completely unwound state, and the value does not re-zero with a complete rotation. The arctan $2(\cdot)$ solution in Eq. (4.28) always provides the minimally wound tilt angle solution, with an orientation for optimal flight efficiency. For the rotor force, the solution in Eq. (4.28) minimizes the vector of rotor force in the least squares sense for minimal overall rotor effort. The attraction velocity to this force vector essentially acts as the same regularization term.

Desired differential inputs from Eq. (4.43) are used in Eq. (4.40) to compute the differential solution. Tilt angle velocities are fed through a saturation function such that $|\dot{\alpha}_i| \leq \dot{\alpha}_{\max}$, and the resulting $\dot{\boldsymbol{u}}$ is integrated with the previous inputs $\boldsymbol{u}_{\text{prev}}$ to produce the final actuator commands.

Through this formulation, kinematic singularities are implicitly handled by expressing control inputs at the differential level, and including the desired unwound state to prevent drift. However, this formulation does not have guarantees that the desired force is met, since actuation constraints are not considered in the problem formulation. Tuning of the gains $\boldsymbol{W}$, v_ω, and v_α becomes very important for the overall system performance.

4.1.5 Rotor Force Mapping

In our implementation of rotary-wing actuators in this work (see Chap. 5), the real actuators take a PWM command as input. At the controller level, we assume that they take a direct thrust command, so we handle the subsequent translation from desired thrust to commanded PWM or rotor speed in a separate module. A block diagram of this module is shown in Fig. 4.8.

Based on the rotor models presented in Sect. 3.1.2, the square of the propeller's angular velocity should relate well with force, but we do not command this velocity directly. Instead, we establish an empirical relationship between the desired rotor angular velocity, $\omega_{r_i}^*$, and PWM commands, $\boldsymbol{u}_{\text{PWM},i}$, as a function of the battery voltage, V_{batt}.

$$\omega_{r_i}^* = \frac{1}{c_f}\text{sign}\left(f_{r_i}^*\right)\sqrt{|f_{r_i}^*|} \qquad \bar{\omega}_{r_i}^* = \frac{\omega_{r_i}^*}{\omega_{r_{\max}}} \qquad \forall i \in 1 \ldots n_r \tag{4.44}$$

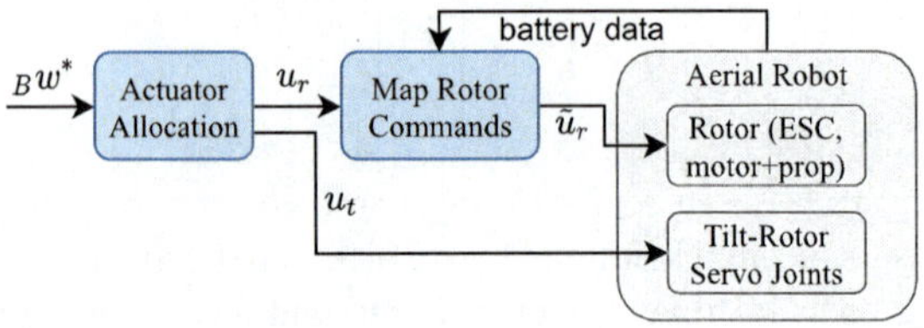

Fig. 4.8 Rotor command mapping transforms force commands to PWM inputs for the electronic speed controllers (ESCs) as a function of the measured battery voltage

$$u_{\text{PWM},i} = \frac{c_{V0} + c_{V1}\bar{\omega}^*_{r_i} + c_{V2}(\bar{\omega}^*_{r_i})^2}{V_{\text{batt}}} \tag{4.45}$$

In Eqs. (4.44) and (4.45), c_{V0}, c_{V1}, c_{V2} are scalar constants for a quadratic fit of empirical data, and $\omega_{r_{\max}}$ is the maximum rotor angular velocity that approximates the linear range, for positive force thrust or invertible thrust with symmetric limits ($\omega_{r_{\max}} = -\omega_{r_{\min}}$). Note that for invertible rotors, special considerations should be taken in the zero-force transition zone, as proposed in [9].

4.2 Interaction Control

In this section, we extend the concept of a free-flight base wrench controller from Sect. 4.1.2 to include contact with the environment. Starting with the description of an *external wrench estimator*, we then develop an *axis-selective impedance control* framework for interaction. We extend this impedance control formulation to exhibit variable compliance in the direction of contact depending on distance sensor feedback. *Direct force control* is then integrated in a hybrid framework with variable impedance control, to incorporate commanded forces and a confidence factor for attempting an interaction task.

4.2.1 *External Wrench Estimation*

To account for the influence of contact forces, we model external contact with a wrench estimator using a generalized momentum approach. Our implementation follows methods described in [10, 11]. Based on preliminary interaction modelling in Sect. 3.3, we integrate the change in momentum over time from Eq. (3.53), and take the difference from the measured system momentum $\boldsymbol{M}\boldsymbol{\nu}_B$ according to the conservation of momentum law. Applying a gain to this residual, we arrive at a new external wrench estimate, $\hat{\boldsymbol{w}}_{\text{ext}}$.

$$\hat{\boldsymbol{w}}_{\text{ext}} = \boldsymbol{K}_I \left(\boldsymbol{M}\boldsymbol{\nu}_B - \int \left(\boldsymbol{C}^\top \boldsymbol{\nu}_B - g + \hat{\boldsymbol{w}}_{\text{ext}} + \boldsymbol{w}_{\text{act}} \right) dt \right) \tag{4.46}$$

The positive-definite diagonal observer matrix $\boldsymbol{K}_I \in \mathbb{R}^{6\times 6}$ acts as an estimator gain. Differentiating Eq. (4.46), a first-order low-pass filtered estimate $\hat{\boldsymbol{w}}_{\text{ext}}$ of the external wrench $\boldsymbol{w}_{\text{ext}}$ is obtained:

$$\dot{\hat{\boldsymbol{w}}}_{\text{ext}} = \boldsymbol{K}_I (\boldsymbol{w}_{\text{ext}} - \hat{\boldsymbol{w}}_{\text{ext}}). \tag{4.47}$$

This solution for a wrench estimator does not require the use of angular acceleration terms which would have to be differentiated from gyroscope angular velocity measurements. The choice of gain $\boldsymbol{K}_I$ influences how aggressively the wrench estimate converges, limited by model accuracy and state estimation noise.

4.2.2 Impedance Control

We shape the interaction behaviour of our system with an impedance controller, using the knowledge of an external wrench to simulate desired system dynamics. The impedance control framework operates as a base-wrench controller, and is therefore platform agnostic, coupled with a morphology-specific actuator allocation module.

A block diagram of the complete control framework in Fig. 4.9 shows the relationship between the impedance control module and the rest of the autonomous system.

Impedance control indirectly regulates a wrench exerted by the system on its environment, without the added complexity of contact detection or controller switching. The same controller can account for interaction, as well as for stable flight in free space. We can take advantage of the system's full actuation to implement an impedance controller with selective apparent inertia, to reject disturbances in some directions while exhibiting compliant behaviour in others, as illustrated in Fig. 4.10.

Our implementation is based on fundamental methods [12], extending to a new application for omnidirectional aerial robots. To begin, we choose the desired closed-loop dynamics of the system to be

$$\boldsymbol{M}_v\dot{\boldsymbol{\nu}} + \boldsymbol{D}_v\boldsymbol{e}_\nu + \boldsymbol{K}_v\boldsymbol{e}_x = \boldsymbol{w}_{\text{ext}}, \tag{4.48}$$

where $\boldsymbol{M}_v$, $\boldsymbol{D}_v$, and $\boldsymbol{K}_v \in \mathbb{R}^{6\times6}$ are positive-definite matrices representing the desired virtual inertia, damping, and stiffness of the system. Pose and twist errors are represented by $\boldsymbol{e}_x$ and $\boldsymbol{e}_\nu \in \mathbb{R}^6$, respectively, corresponding with stacked geometric translational and rotational error from Eq. (4.8).

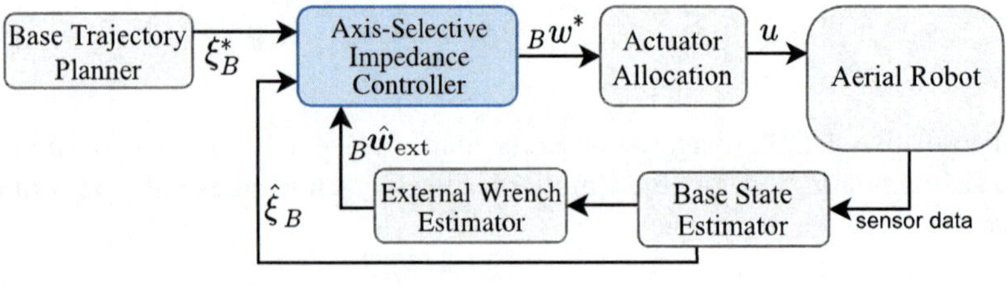

Fig. 4.9 Block diagram of axis-selective impedance control framework, with system state, $\boldsymbol{\xi}$, and actuator commands, $\boldsymbol{u}$

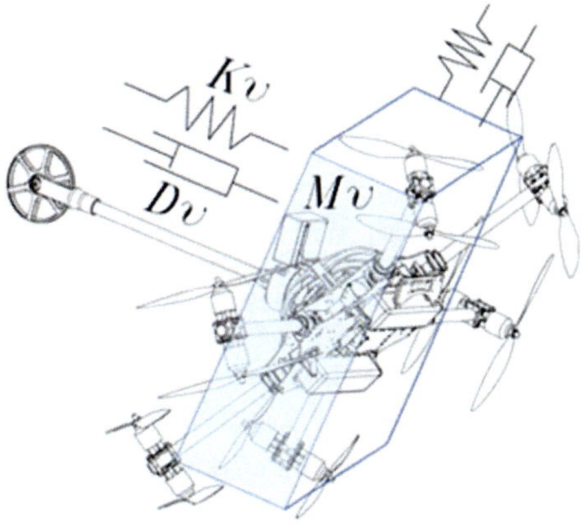

Fig. 4.10 Visualization of the virtual mass effect, where a box overlaid on a system diagram represents axis-selective resistance or compliance to external forces

The system dynamics in Eq. (3.48) and desired dynamics in Eq. (4.48) are then combined by substituting the stacked acceleration vector, $\dot{\boldsymbol{\nu}}$, using the estimate from Eq. (4.46) for the external wrench, $\hat{\boldsymbol{w}}_{\text{ext}}$. We then rearrange the equations to solve for the wrench control input, $\boldsymbol{w}_{\text{act}}$.

$$\boldsymbol{w}_{\text{act}} = (\boldsymbol{M}\boldsymbol{M}_v^{-1} - \mathbb{1}_6)\hat{\boldsymbol{w}}_{\text{ext}} - \boldsymbol{M}\boldsymbol{M}_v^{-1}(\boldsymbol{D}_v\boldsymbol{e}_\nu + \boldsymbol{K}_v\boldsymbol{e}_x) + \boldsymbol{C}\boldsymbol{\nu} + g \tag{4.49}$$

Since the stiffness and damping properties of interaction depend highly on the apparent inertia, we first normalize these matrices with respect to the system mass as $\tilde{\boldsymbol{M}}_v = \boldsymbol{M}^{-1}\boldsymbol{M}_v$, then express stiffness and damping as

$$\tilde{\boldsymbol{D}}_v = \tilde{\boldsymbol{M}}_v^{-1}\boldsymbol{D}_v \qquad \tilde{\boldsymbol{K}}_v = \tilde{\boldsymbol{M}}_v^{-1}\boldsymbol{K}_v \tag{4.50}$$

The virtual dynamics which may be axis-specific can be rotated into any desired frame. For point contact aerial interaction tasks, we choose alignment with the end-effector frame, using $\boldsymbol{R} = \text{blockdiag}\{\boldsymbol{R}_{BE}, \boldsymbol{R}_{BE}\}$. We then rewrite Eq. (4.49) as

$$\boldsymbol{w}_{\text{act}} = (\boldsymbol{R}^\top\tilde{\boldsymbol{M}}_v^{-1}\boldsymbol{R} - \mathbb{1}_6)\hat{\boldsymbol{w}}_{\text{ext}} - \tilde{\boldsymbol{D}}_v\boldsymbol{e}_\nu - \tilde{\boldsymbol{K}}_v\boldsymbol{e}_x + \boldsymbol{C}\boldsymbol{\nu} + g. \tag{4.51}$$

Using a rigidly attached end-effector as the simplest solution for aerial interaction, the relative transformation $\boldsymbol{T}_{BE}$ is constant. We choose to exhibit selective compliance along the $\boldsymbol{e}_z^E$ axis, assuming that the planner aligns this axis with the local contact surface normal. A high virtual mass is set in the orthogonal plane, and orientation, for a stronger resistance to aerodynamic disturbances.

To adapt the PID wrench controller from Sect. 4.1.2 to an impedance framework, we realize that Eq. (4.51) is equivalent to our PID control, with a wrench estimate term in place of an integrator. We can therefore adopt the same gains $\tilde{\boldsymbol{K}}_v = \boldsymbol{K}_p$ and $\tilde{\boldsymbol{D}}_v = \boldsymbol{K}_d$, and need only to choose the ratio of the desired system inertia matrix as a function of the model inertia, expressed in the end-effector frame as

$$\tilde{\boldsymbol{M}}_v = \mathrm{diag}\{\left[m_{v,\mathrm{free}}\ m_{v,\mathrm{free}}\ m_{v,E}\ \mathcal{I}_v\ \mathcal{I}_v\ \mathcal{I}_v\right]^\top\}, \tag{4.52}$$

where scalar values $m_{v,\mathrm{free}}$ and $\mathcal{I}_v$ are typically set to be greater than 1 to exhibit resistance to estimated external forces, and $m_{v,E}$ is set less than 1 to exhibit compliance in interaction.

Since model error is lumped into the external wrench estimate, this method has the drawback that the compliant interaction axis is also compliant to model error and aerodynamic disturbances, i.e. , the system will allow greater tracking error along this axis, even in free flight.

Variable Axis-Selective Impedance Control

To address the free-flight problem that axis-selective impedance control is compliant to model error and disturbances, we vary the value for $m_{v,E}$ along the intended contact axis. In this way, we can have a high virtual mass along this axis in free flight while maintaining compliance during physical interaction tasks.

We use distance measurement d_t from the tool frame to the nearest locally observed surface that intersects $\boldsymbol{e}_z^E$. In free flight, when surfaces are farther than a threshold $d_{\max}$, the virtual mass along the end-effector axis is set to $m_{v,\mathrm{free}}$, to reject disturbances like any other axis. Within a range from $d_{\max}$ to $d_{\min}$ from the surface, that virtual mass reduces to $m_{v,\mathrm{wall}} < 1$ according to a cosine function to exhibit interaction compliance. The relationship between d_t and $m_{v,\mathrm{wall}}$ is shown in Fig. 4.11. We calculate the virtual mass in the end-effector direction, $m_{v,E}$, as follows:

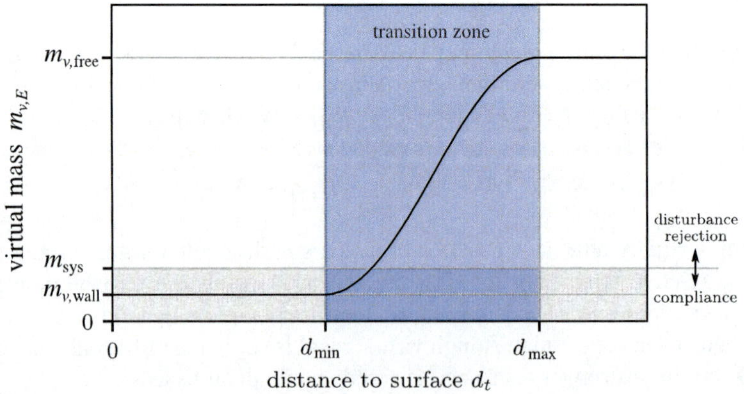

Fig. 4.11 Variable impedance in the $\boldsymbol{e}_z^E$ direction as a function of wall distance d_t, with transition zone shown in blue, where m_{sys} is the true system mass

$$c_{v,E} = \begin{cases} 1, & \text{if } d_t \leq d_{\min} \\ 0.5(1 + \cos(\frac{d_t - d_{\min}}{d_{\max} - d_{\min}}\pi)), & \text{if } d_{\min} < d_t \leq d_{\max} \\ 0, & \text{otherwise} \end{cases} \tag{4.53}$$

$$m_{v,E} = c_{v,E}(m_{v,\text{free}} - m_{v,\text{wall}}) + m_{v,\text{wall}}. \tag{4.54}$$

The complete virtual mass matrix is then constructed as before in Eq. (4.52).

4.2.3 Hybrid Force-Impedance Interaction Control

In this section we develop an interaction control scheme that incorporates some awareness of its task and environment, adjusting the controller action accordingly in case of unexpected interaction or missing contact surfaces. This hybrid method incorporates distance sensing from variable axis-selective impedance control, as well as tactile sensing with a force sensor at the end-effector, in an effort to improve task performance while also increasing robustness of the system. A block diagram summarized the controller and its interfacing components in Fig. 4.12.

The momentum-based external wrench contains an accumulation of force and torque unrelated to the point of interaction, and in many cases cannot be used for direct force tracking. With direct multi-axis force sensing at the end-effector, contact forces can be isolated from remaining forces caused by model error and aerodynamic disturbances, such that desired forces at the end-effector can be resolved explicitly. The following content describes force control at the end-effector and integration with variable axis-selective impedance control in a hybrid scheme, based on the collaborative work published in [13].

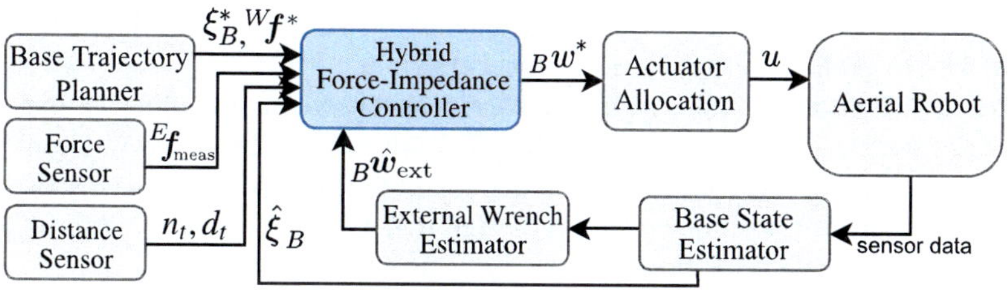

Fig. 4.12 Block diagram of hybrid force-impedance control framework, combining direct force control with variable axis-selective impedance control (ASIC). The primary difference from Fig. 4.9 is the integration of direct force f_{meas} and distance sensing (d_t) in the controller

Direct Force Control

In this framework, the trajectory sent from the planner contains an additional field, representing the desired world frame force vector ${}^{W}\boldsymbol{f}^{*}(t) \in \mathbb{R}^{3}$ acting on the environment for a particular interaction task. We can directly close the loop on force control using feedback from an integrated sensor at the end-effector, which produces a force reading, ${}^{E}\boldsymbol{f}_{\text{meas}}$, in the end-effector frame $\mathcal{F}_{E}$. Since the measured force represents the reaction to the exerted contact forces, this term is summed with the desired force to calculate the force tracking error in $\mathcal{F}_{B}$.

$$\boldsymbol{e}_{f} = \boldsymbol{R}_{BE}\,{}^{E}\boldsymbol{f}_{\text{meas}} + \boldsymbol{R}_{WB}^{\top}\,{}^{W}\boldsymbol{f}^{*} \tag{4.55}$$

A proportional-integral (PI) control scheme with a feed-forward force term is implemented to track the given reference force, closing a feedback loop on the force tracking error $\boldsymbol{e}_{f}$ from Eq. (4.55):

$$\boldsymbol{f}_{\text{dir}} = \frac{1}{m}\left(\boldsymbol{R}_{WB}^{\top}\,{}^{W}\boldsymbol{f}^{*} - k_{f,p}\boldsymbol{e}_{f} - k_{f,i}\int \boldsymbol{e}_{f}\text{dt}\right), \tag{4.56}$$

where $k_{f,p}$ and $k_{f,i}$ are positive scalar gains.

Hybrid Control

The proposed sensor-based hybrid controller incorporates the idea of interaction task awareness, where a zero force trajectory sent by the planner indicates free flight, and a non-zero force indicates intended interaction. Force control is only attempted when it is explicitly communicated by the planner, and tracking of the pose trajectory for all axes is performed at all times. At points of interaction, the desired tool trajectory should lie on the interaction surface, while the additional desired force vector determines interaction behaviour.

We define a contact task *confidence factor* λ, representing confidence that the reference force can exist at the target location without deviating too far from the planned trajectory. This confidence is a function of the perceived surface distance d_t and the end-effector position error $\boldsymbol{e}_t$ projected in the direction of desired force. We compute the projected end-effector error, e_{t,f^*}, as

$$\boldsymbol{e}_t = {}^{W}\boldsymbol{p}_{E} - {}^{W}\boldsymbol{p}_{E}^{*} \in \mathbb{R}^{3} \tag{4.57}$$

$$e_{t,f^*} = \frac{\boldsymbol{e}_t \cdot {}^{W}\boldsymbol{f}^{*}}{||\,{}^{W}\boldsymbol{f}^{*}||}. \tag{4.58}$$

The confidence factor is smoothly transitioned with a first-order filter with coefficient c_{λ} to avoid step inputs at the start or end of a non-zero desired force. We arrive at the computation of the confidence factor λ_k which represents λ at time step k.

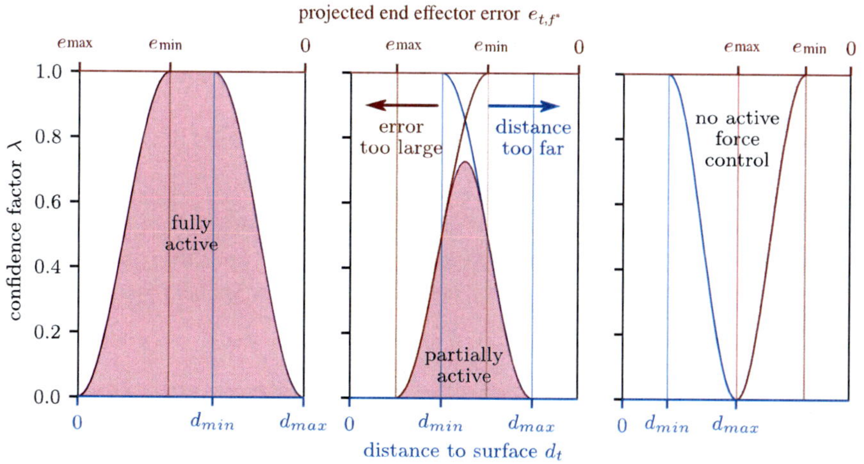

Fig. 4.13 Confidence factor λ as a function of wall distance d_E, and projected tool error e_{E,f^*}. The solid area indicates the proportion of force control used

$$\lambda_d = \begin{cases} 1, & \text{if } d_t \leq d_{\min} \\ 0.5(1 + \cos(\frac{d_t - d_{\min}}{d_{\max} - d_{\min}}\pi)), & \text{if } d_{\min} < d_t \leq d_{\max} \\ 0, & \text{otherwise} \end{cases} \tag{4.59}$$

$$\lambda_e = \begin{cases} 1, & \text{if } e_{t,f^*} \leq e_{\min} \\ 0.5(1 + \cos(\frac{e_{t,f^*} - e_{\min}}{e_{\max} - e_{\min}}\pi)), & \text{if } e_{\min} < e_{t,f^*} \leq e_{\max} \\ 0, & \text{otherwise} \end{cases} \tag{4.60}$$

$$\lambda_k = \begin{cases} c_\lambda \lambda_d \lambda_e + (1 - c_\lambda)\lambda_{k-1}, & \text{if } ||\boldsymbol{f}^*|| > 0 \\ 0, & \text{otherwise} \end{cases} \tag{4.61}$$

The behaviour of the combined force and impedance control is shown in Fig. 4.13. In the nominal case, the planner commands a path in free flight which the controller can achieve, and a desired force is commanded only when the sensed distance d_t and projected tool error $\boldsymbol{e}_{t,f_*}$ are small. In the case where the set point is behind the wall, the controller uses compliant impedance control in the direction of the end-effector to perform its task as well as possible. When a force is then commanded, λ is 1 and force control is fully active. When the set point is in front of the wall between $d_{\min}$ and $d_{\max}$, there is a transition phase where λ is between 1 and 0, a compromise between trying to achieve force control and maintain trajectory tracking. In the case where the wall is not sensed within $d_{\max}$, λ is 0 and no force control is attempted.

A selection matrix, $\boldsymbol{\Lambda} \in \mathbb{R}^{3\times3}$, orients the confidence factor λ from Eq. (4.61) in the direction of desired force ${}^{B}\boldsymbol{f}^*$ using the rotation $\boldsymbol{R}_{f^* e_z}$, and is constructed as follows:

$$\boldsymbol{\Lambda} = \boldsymbol{R}_{f^* e_z} \begin{bmatrix} 0 & 0 & 0 \\ 0 & 0 & 0 \\ 0 & 0 & \lambda \end{bmatrix} \boldsymbol{R}_{f^* e_z}^{\top} \in \mathbb{R}^{3\times 3} \tag{4.62}$$

$$\widetilde{\boldsymbol{\Lambda}} = \text{blockdiag}\{\boldsymbol{\Lambda}, \boldsymbol{0}_{3\times 3}\} \in \mathbb{R}^{6\times 6} \tag{4.63}$$

The selection matrix is extended to 6 DOF as $\widetilde{\boldsymbol{\Lambda}} \in \mathbb{R}^{6\times 6}$ by padding with zeros since we have a point end-effector, but this could be extended in the future to include interaction torques. In the present implementation, the matrix positively selects direct force control commands, producing an actuation wrench component associated with direct force control:

$$\boldsymbol{w}_{\text{dir}} = \begin{bmatrix} \boldsymbol{\Lambda} \boldsymbol{f}_{\text{dir}} \\ \boldsymbol{0}_{3\times 1} \end{bmatrix} \in \mathbb{R}^{6}. \tag{4.64}$$

Using $\tilde{\boldsymbol{M}}_v$ from Eq. (4.54) we compute the variable axis-selective impedance control command in the absence of dynamic terms as

$$\boldsymbol{w}_{\text{imp}} = (\mathbb{1}_6 - \widetilde{\boldsymbol{\Lambda}})(\boldsymbol{R}^{\top} \tilde{\boldsymbol{M}}_v^{-1} \boldsymbol{R} - \mathbb{1}_6)\hat{\boldsymbol{w}}_{\text{ext}} - \tilde{\boldsymbol{D}}_v \boldsymbol{e}_{\nu} - \tilde{\boldsymbol{K}}_v \boldsymbol{e}_x, \tag{4.65}$$

where the selection matrix counterpart $(\mathbb{1}_6 - \widetilde{\boldsymbol{\Lambda}})$ negatively selects the component of the momentum-based wrench estimate in the direction of desired interaction.

The two control components are then combined and compensated for nonlinear dynamic effects and gravity. Since the system COM is offset from the system's geometric center due to an extended end-effector, we use a feed-forward term to compensate for the resulting torque. The final resulting wrench command is shown as a block diagram in Fig. 4.14, and computed as follows:

$$\boldsymbol{w}^*_{\text{cmd}} = \boldsymbol{w}_{\text{dir}} + \boldsymbol{w}_{\text{imp}} + \boldsymbol{C}\boldsymbol{\nu} + g \tag{4.66}$$

$$\boldsymbol{w}_{\otimes} = \begin{bmatrix} \boldsymbol{0}_{3\times 1} \\ {}^{B}\boldsymbol{p}_{\otimes} \times \boldsymbol{\tau}_{\text{cmd}} \end{bmatrix} \tag{4.67}$$

$$\boldsymbol{w}_{\text{cmd}} = \boldsymbol{w}^*_{\text{cmd}} + \boldsymbol{w}_{\otimes}. \tag{4.68}$$

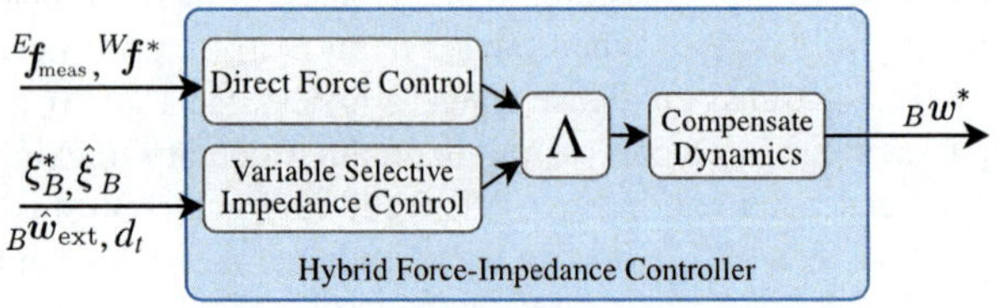

Fig. 4.14 Hybrid force-impedance control detail, with impedance control varying apparent mass along the $\boldsymbol{e}_z^E$ axis as a function of the sensed distance d_t. Direct force control is added selectively as a function of d_t and the end-effector error $\boldsymbol{e}_t$

The resulting actuation command from Eq. (4.68) is then sent to an actuator allocation module as described in Sect. 4.1.3 to assign corresponding commands to the system's actuators.

4.3 Macro-Micro Control for Redundant Aerial Manipulators

In this section we describe the control of a parallel manipulator mounted to an omnidirectional flying base in a decoupled control scheme, integrating the dynamic reaction forces of the manipulator as a feed-forward term in the base controller. Our aim with this controller is to make use of the macro-micro model for dynamic end-effector trajectory tracking with high precision, despite positioning errors and disturbances applied to the flying base. Interaction forces at the end-effector can be incorporated by extending this formulation, but this remains a topic of future work. We consider the case in which the reference trajectories for the end-effector position, ${}^{W}\boldsymbol{p}_E{}^*(t)$, and the flying base position and attitude, ${}^{W}\boldsymbol{p}_B^*(t)$ and $\boldsymbol{R}_{WB}^*(t)$ are provided by a planning module. We recall that, because of the delta's 3 translational DOFs, the attitude of the end-effector and of the flying base are the same.

Under these conditions, we aim to achieve the following decoupled control goals:

- Manipulator action minimizes the end-effector position error given the current base pose.
- Flying base action minimizes its own position and attitude errors.

Given the two control objectives, and employing the dynamic model of the system from Sect. 3.4, we implement a standard inverse dynamic plus linear control action. This also allows for proper compensation of the dynamic coupling effects between the flying base and robot arm.

Due to weight limitations, robotic arms for aerial manipulators are normally realized with servomotors which take position or velocity control inputs, instead of direct joint torque commands. The flying base, however, can directly control an aerodynamic actuation wrench. The following presents how we handle the control between the manipulator and the flying base, exploiting the manipulator's dynamic model to compensate dynamic coupling effects.

4.3.1 Delta Manipulator Control

We briefly describe inverse kinematic control approaches for position- and velocity-controlled servomotor joints, as seen in state-of-the-art parallel manipulators.

Position Control

If the arm joints are controlled in position, the servomotors internally run a PID controller to track a desired position reference input $\boldsymbol{q}_{m^*}$. Given the current base pose, ${}^W\boldsymbol{p}_B$ and $\boldsymbol{R}_{WB}$, and desired end-effector position ${}^W\boldsymbol{p}_E{}^*$, we can compute the joint references by inverting the kinematic relation from Eq. (3.56):

$$\boldsymbol{q}_{m^*} = h_E^{-1}({}^W\boldsymbol{p}_B, \boldsymbol{R}_{WB}, {}^W\boldsymbol{p}_E{}^*). \tag{4.69}$$

Velocity Control

If the arm joints are controlled in velocity, the servomotors internally run a PI controller, tracking a desired velocity reference input $\dot{\boldsymbol{q}}_{m^*}$. Given the current base pose, $({}^W\boldsymbol{p}_B, \boldsymbol{R}_{WB})$, base velocities, $({}^W\boldsymbol{v}_B, \boldsymbol{\omega}_B)$, and desired end-effector position and velocity $({}^W\boldsymbol{p}_E{}^*, {}^W\boldsymbol{v}_E{}^*)$, we can compute the joint velocity references by proportional controller plus a feed-forward term in an outer loop to the servomotor PI controller by using the manipulator Jacobian, $\boldsymbol{J}_D$ as:

$$\dot{\boldsymbol{q}}_{m^*} = \boldsymbol{J}_D(\boldsymbol{q}_m)^{-1}\,{}^W\boldsymbol{v}_{EB}{}^* + \boldsymbol{K}_P^D(\boldsymbol{q}_{m^*} - \boldsymbol{q}_m), \tag{4.70}$$

where $\boldsymbol{K}_P^D \in \mathbb{R}^{3\times 3}$ is a positive definite gain matrix. The desired end-effector linear velocity with respect to the current $\mathcal{F}_B$ is computed as

$${}^W\boldsymbol{v}_{EB}{}^* = {}^W\boldsymbol{v}_E{}^* - \boldsymbol{J}_{PB}(\boldsymbol{q}_m)\,{}^W\boldsymbol{v}_B - \boldsymbol{J}_{RB}(\boldsymbol{q}_m)\boldsymbol{\omega}_B, \tag{4.71}$$

and $\boldsymbol{q}_{m^*}$ is computed as in Eq. (4.69).

In ideal conditions, velocity control is the preferred option due to the presence of a feed-forward term. In our real implementation, the selected servomotors have limited read/write frequency which subsequently limits the frequency of the outer controller Eq. (4.70). The maximum value of $\boldsymbol{K}_P^D$ must also be lowered, resulting in reduced tracking performance. Since the frequency of the outer loop controller is much slower than the internal servo controller, we achieve better performance in practice with a position control implementation, relying on the high-rate internal position controller of the servo.

Workspace Projection

The real manipulator is subject to physical constraints, and requires avoiding singular configurations mentioned in Sect. 3.4.1. We handle infeasible commanded positions in the inverse kinematic position control formulation by first projecting target points onto the known feasible manipulator workspace.

Figure 4.15a shows the limitations for an individual chain of the parallel structure, offset by the end-effector plate radius such that the end of the distal link corresponds

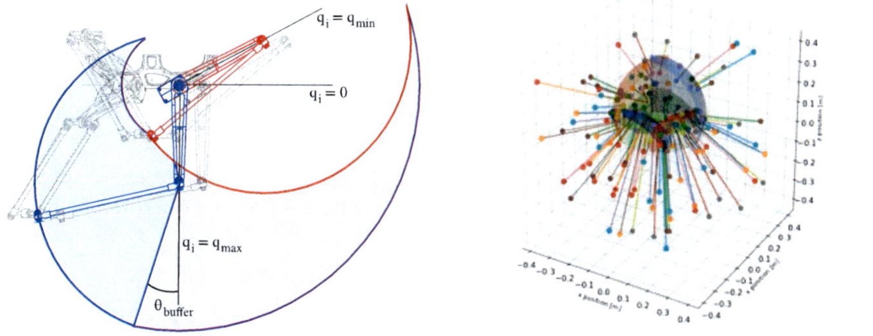

(a) Workspace visualisation of a single kinematic chain i offset by the end-effector plate radius.

(b) Projection of randomly selected infeasible points onto the singularity-free feasible workspace in $\mathcal{F}_D$.

Fig. 4.15 Delta manipulator workspace diagrams

with the end-effector point. For joint position q with a horizontal zero reference, the minimum and maximum positions $q_{\min}$ and $q_{\max}$ are due to interference with the mounting base hardware. Rotation of the distal link relative to the proximal link is limited by physical intersection with the distal link and limits of the parallelogram joints. Due to a serial chain singularity that occurs when the two links align relative to the actuated axis, the distal link is constrained to maintain a buffer angle θ_{buffer} from the singularity, preventing elbow inversion. The workspace of a single chain is the area covered by the near-hemispherical surface swept from $q_{\min}$ to $q_{\max}$. The complete end-effector workspace is the intersection of these three volumes when patterned around the $\boldsymbol{e}_z^D$ axis.

Since reference positions may shift outside of the delta workspace due to disturbance of the floating base, we identify infeasible points and project them back onto the feasible space. The delta workspace expressed in $\mathcal{F}_D$ can be represented by 3 spherical shells corresponding to $q_{\max}$ as the outer bounds, centered at $\boldsymbol{p}_{ei}(q_{\max})$, and those corresponding to $q_{\min}$ as the inner bounds, $\boldsymbol{p}_{ei}(q_{\min})$. Upper and lower bounding spheres have radii $l_D - \epsilon$ and $l_D + \epsilon$ respectively, where ϵ is a small buffer to ensure the projected point lies within the workspace. We further know $z_{\min}$, the minimum feasible z coordinate in $\mathcal{F}_D$. For any point candidates for end-effector position $\boldsymbol{p}_E^* = [x^*, y^*, z^*]^\top$, we perform the following projection steps:

1. Project to min height if below: $z^* = \max(z_{\min}, z^*)$,
2. Determine region i of workspace in the z-plane.
3. If $||\boldsymbol{p}_E^* - \boldsymbol{p}_{ei}(q_{\max})|| > l_D - \epsilon$, project to outer sphere.
4. If $||\boldsymbol{p}_E^* - \boldsymbol{p}_{ei}(q_{\min})|| < l_D + \epsilon$, project to inner sphere.

Figure 4.15b visualizes 100 random points, where kinematic violations are projected back onto the workspace, bounded by the intersecting surfaces. In the case of velocity control, the target velocity is set to zero if the point is invalid before projection.

4.3.2 Feed Forward Dynamics

Based on the dynamic model of the floating base parallel manipulator derived in Sect. 3.4.3, we construct a base wrench control formulation to compensate reaction forces from a fast-moving manipulator.

From the projected Newton-Euler dynamic formulation, we stop at Eq. (3.63). The system dynamics of the base wrench in its own frame are then written as

$$\boldsymbol{M}(\boldsymbol{q}_b)\dot{\boldsymbol{\nu}}_B + \boldsymbol{C}(\boldsymbol{q}_b, \boldsymbol{\nu}_B)\boldsymbol{\nu}_B + g(\boldsymbol{q}_b) + {}^B\boldsymbol{w}_\Delta = {}^B\boldsymbol{w}_{\text{act}} + {}^B\boldsymbol{w}_{\text{ext}}, \tag{4.72}$$

where $\mathcal{F}_{P_{ai}}$ is the frame corresponding to the ith joint frame for $i \in \{1 \ldots 3\}$ actuated joints.

The final wrench control command is generated by combining Eq. (4.72) with a single body base wrench controller such as the PID free-flight control formulation in Eq. (4.14), assuming external forces are small enough to be treated as disturbances.

$${}^B\boldsymbol{w}^*_{\text{act}} = \boldsymbol{M}(\boldsymbol{q}_b)\,{}^B\dot{\boldsymbol{\nu}}_{\text{PID}} + \boldsymbol{C}(\boldsymbol{q}_b, \boldsymbol{\nu}_B)\boldsymbol{\nu}_B + g(\boldsymbol{q}_b) + {}^B\boldsymbol{w}_\Delta \tag{4.73}$$

The control wrench generated in Eq. (4.73) is then fed through a direct actuator allocation module to generate tilt-rotor actuator inputs. The block diagram of the control wrench implementation in Eq. (4.73) and its connecting modules are shown in Fig. 4.16.

We note that while manipulator dynamics compensation only receives feedback in $\boldsymbol{q}_m$ and $\dot{\boldsymbol{q}}_m$ from the system, although $\ddot{\boldsymbol{q}}_m$ is also required for the dynamic computation. In practice, we generate the acceleration terms by differentiating the joint velocities and applying a first-order filter to remove noise. Delay from the filter adds some error to the estimated reaction force, which grows with more dynamic end-effector movement. We also note the implicit assumption that the base controller can immediately compensate high-frequency reaction forces from the manipulator,

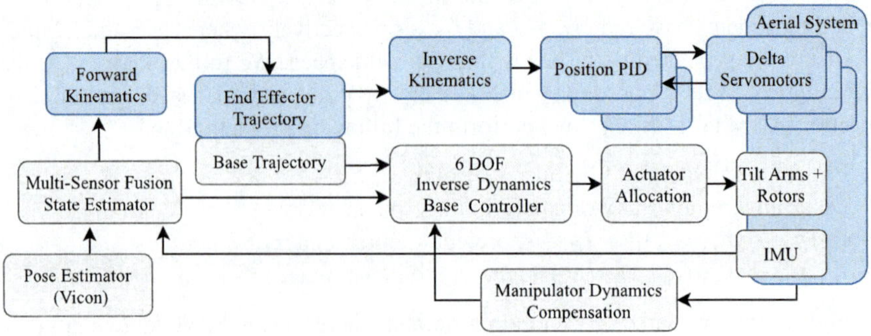

Fig. 4.16 Block diagram for end-effector control of a redundant aerial manipulator, where dynamic reaction forces from the manipulator are compensated in as independent base controller with a feed-forward term

which in the case of our current tilt-rotor actuator allocation methods is not necessarily met. The consequences of this semi-coupled controller implementation will become clear in experiments in Sect. 6.4, and motivate the need for a whole-body controller of the complete tilt-rotor and manipulator system.

References

1. Bouabdallah, S., Siegwart, R.: Full control of a quadrotor. In: 2007 IEEE/RSJ International Conference on Intelligent Robots and Systems, pp. 153–158. IEEE (2007)
2. Lee, T., Leok, M., McClamroch, N.H.: Geometric tracking control of a quadrotor UAV on se (3). In: 49th IEEE Conference on Decision and Control (CDC), pp 5420–5425. IEEE
3. Allenspach, M., Bodie, K., Brunner, M., Rinsoz, L., Taylor, Z., Kamel, M., Siegwart, R., Nieto, J.: Design and optimal control of a tiltrotor micro-aerial vehicle for efficient omnidirectional flight. Int J Robot Res **39**(10–11), 1305–1325 (2020)
4. Brunner, M., Bodie, K., Kamel, M., Pantic, M., Zhang, W., Nieto, J., Siegwart., R.: Trajectory tracking nonlinear model predictive control for an overactuated mav. In: 2020 IEEE International Conference on Robotics and Automation (ICRA), pp. 5342–5348. IEEE
5. Zhang, W., Brunner, M., Ott, L., Kamel, M., Siegwart, R., Nieto, J.: Learning dynamics for improving control of overactuated flying systems. IEEE Robot Autom Lett **5**(4), 5283–5290 (2020)
6. Kamel, M., Verling, S., Elkhatib, O., Sprecher, C., Wulkop, P., Taylor, Z., Siegwart, R., Gilitschenski, I.: The Voliro omniorientational hexacopter: an agile and maneuverable Tiltable-Rotor aerial vehicle. IEEE Robot Autom Mag **25**(4), 34–44 (2018)
7. Falconi, R., Melchiorri, C.: Dynamic model and control of an over-actuated quadrotor UAV. IFAC Proc. Vol. **45**(22), 192–197 (2012)
8. Wampler, C.W.: Manipulator inverse kinematic solutions based on vector formulations and damped least-squares methods. IEEE Trans Syst Man Cybernet **16**(1), 93–101 (1986)
9. Park, S., Lee, J., Ahn, J., Kim, M., Her, J., Yang, G.-H., Lee, D.: ODAR: aerial manipulation platform enabling omnidirectional wrench generation. IEEE/ASME Trans Mechatron **23**(4), 1907–1918 (2018)
10. Ruggiero, F., Cacace, J., Sadeghian, H., Lippiello, V.: Impedance control of VToL UAVs with a momentum-based external generalized forces estimator. In: 2014 IEEE International Conference on Robotics and Automation (ICRA), pp. 2093–2099. IEEE
11. Tomić, T., Haddadin, S.: A unified framework for external wrench estimation, interaction control and collision reflexes for flying robots. In: 2014 IEEE/RSJ International Conference on Intelligent Robots and Systems, pp. 4197–4204. IEEE
12. Siciliano, B., Sciavicco, L., Villani, L., Oriolo, G.: Robotics: Modelling, Planning and Control. Springer Science & Business Media (2010)
13. Bodie, K., Brunner, M., Pantic, M., Walser, S., Pfändler, P., Angst, U., Siegwart, R., Nieto, J.: Active interaction force control for contact-based inspection with a fully actuated aerial vehicle. IEEE Trans. Robot. (2020)

Chapter 5
Prototype Design

In this chapter, we bridge the gap between generalized modelling and control frameworks and the realized hardware prototype system on which these methods are implemented. Section 5.1 includes the detailed mechanical design of the omnidirectional tilt-rotor system, Sect. 5.2 discusses electrical connectivity for power and communication, and Sect. 5.3 presents final prototype systems that are used for flight experiments.

5.1 Mechanical Design

The mechanical design of a multi-rotor system is an iterative process. Starting with an initial estimate of the total system mass, we select rotor and propeller combinations that satisfy our design requirements. Tilt-able rotor groups are considered in terms of their inertial properties, and appropriate tilt-arm actuators are selected. Power requirements and desired flight time influence the choice of battery, which is a large part of the overall system mass. Adding the new component masses to the fixed core components, structural estimation and required payload, we arrive at a new system mass, and the iterative process begins again. Figure 5.1 gives an example of this design iteration process.

The scope of mechanical design covers the selection of essential system components and their mechanical integration:

- Rotor groups Sect. 5.1.1.
- Tilting mechanics Sect. 5.1.2.
- The system core Sect. 5.1.3.

In order to interact with the environment, we require a manipulator arm that is capable of transmitting the actuation forces and torques to an end-effector point that

K. Bodie et al., *Omnidirectional Tilt-Rotor Flying Robots for Aerial Physical Interaction*, Springer Tracts in Advanced Robotics 157,
https://doi.org/10.1007/978-3-031-45497-4_5

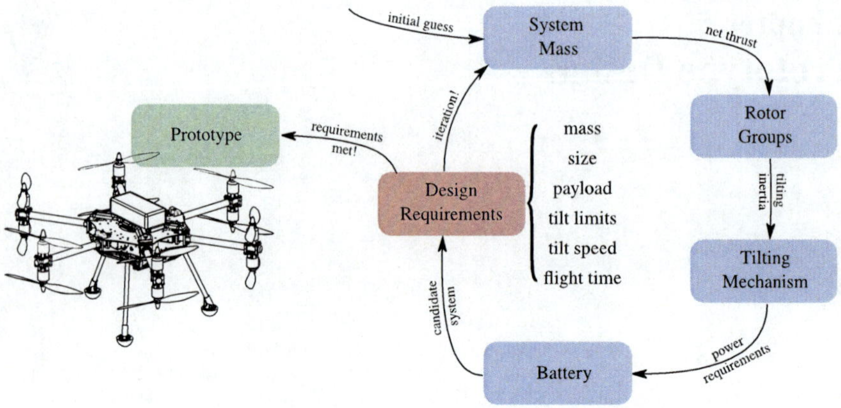

Fig. 5.1 Example design iteration problem for tilt-rotor component selection

is a safe distance away from the system's spinning rotors. The following subsections develop designs for two manipulator strategies:

- Fixed manipulator with various end-effectors Sect. 5.1.4.
- Delta parallel manipulator for macro-micro manipulation Sect. 5.1.5.

5.1.1 Rotor Groups

Rotor groups consisting of a brushless direct current (BLDC) motor, a propeller, and an ESC provide aerodynamic forces to control the vehicle. The net thrust requirement of the system combined with its selected morphology drive the individual rotor thrust requirement. The components are then selected to achieve complementary performance in the desired thrust range.

We start with the flat hexacopter tilt-rotor morphology selected in Sect. 3.2.3, and a total system mass of 5kg. We then calculate the required rotor thrust for sufficiently dynamic flight in the morphology's least efficient rotor configuration (e.g. $\boldsymbol{e}_y^B = \boldsymbol{e}_z^W$). For highly dynamic flight, we choose that the system can accelerate at $|g|$ in all directions, corresponding with a net rotor force that supports $2m|g| = 10$kg. Based on the system morphology, we are interested in evaluating rotor-propeller performance in the following cases:

- Hover on axis $\boldsymbol{e}_z^B \rightarrow 6$ rotor groups at 0.83 kg each.
- Hover on axis $\boldsymbol{e}_y^B \rightarrow 4$ rotor groups at 1.25 kg each.
- Maximum acceleration on axis $\boldsymbol{e}_y^B \rightarrow 4$ rotor groups at 2.5 kg each.

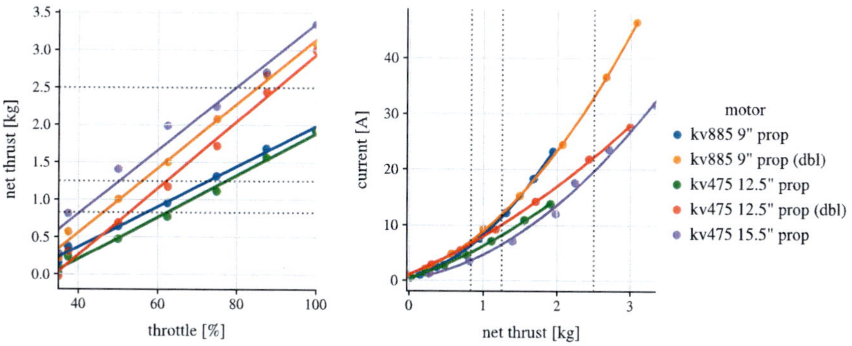

(a) Net thrust per throttle percentage. (b) Current draw per net thrust.

Fig. 5.2 Properties of a single rotor group for two different motors from specifications available at [1], with 6S linear quadratic regulator with integral action (LiPo) battery and propeller combinations as listed in Table 5.1

Table 5.1 Motor and propeller combinations for theoretical evaluation

Motor	Propeller	Single	Double
KDE 2315XF 885Kv	9 × 3in	×	×
KDE 3510XF 475Kv	12.5 × 4.3in	×	×
KDE 3510XF 475Kv	15.5 × 5.3in	×	–

Single Verses Double Rotor Groups

As discussed in Sect. 3.1.3, double rotor groups suffer from efficiency loss due to airflow interaction between co-axial rotors, but have the advantage of decreasing the overall system size. For the desired thrust values determined above, we evaluate the effect of choosing a single or double rotor group for different propeller sizes on the estimated current draw and overall system diameter.

Figure 5.2 compares throttle, net thrust and current draw characteristics for rotor-propeller combinations listed in Table 5.1. Dotted lines correspond to the thrust values that interest us for efficient hover, worst-case hover, and highly dynamic flight. All data are for the case of a 6S LiPo battery (22. V nominal). Throttle values in percentage correspond to PWM commands in the range [0, 1] that are sent to the rotors' ESCs. Presented net thrust values in [kg] represent the thrust given by the manufacturer minus the mass of the rotor group, which is multiplied by 1.6 for double rotor data points to incorporate a 20% efficiency loss. Current values for double rotor groups are simply doubled. Based on empirical data, to remain in a thrust region with a good linear approximation, the maximum output of the rotors is limited to $\approx 80\%$, which we choose for a maximum value.

Observing the relation between net thrust and percent throttle in Fig. 5.2a, the two smaller propeller single groups do not provide sufficient thrust in the highly dynamic

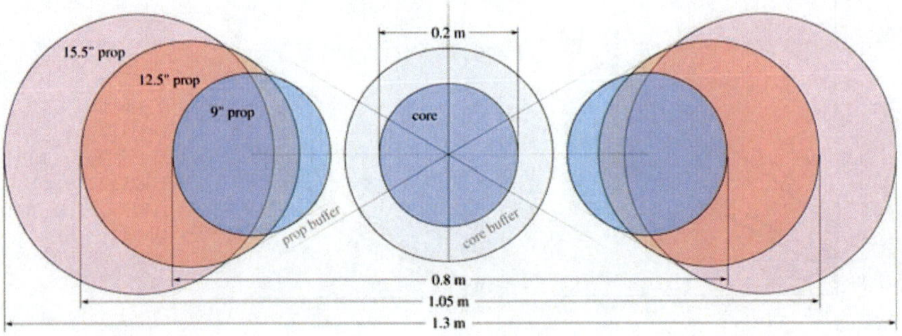

Fig. 5.3 A top view of the hexacopter geometry showing two opposing rotor groups and the system core. The overall system diameter varies with changing propeller diameter

case. Between the remaining groups, the 9in co-axial propeller is the only candidate that can provide the required thrust at 80% throttle after subtraction of the motor mass, though the 12.5in co-axial and 15.5in single come close enough to remain in consideration. We then consider the current draw of these groups in Fig. 5.2b, where the 9in co-axial group now shows the worst performance, demanding 45 A, 60 A and 120 A for the flight cases described above.

An increased propeller size results in a larger increased geometry than the difference in propeller diameters, as seen in Fig. 5.3. Assuming a core diameter of 20cm, and a required aerodynamic buffer zone from the core and neighbouring propellers, the overall system geometry increases substantially with increasing rotor diameter.

Based on the satisfactory thrust performance despite high current-draw, and our desire to keep the overall system size small for indoor testing, we choose to use the smallest (9 in) propeller diameter and corresponding motor listed above in a counter-rotating co-axial configuration. We choose counter-rotating propellers such that drag torques cancel in the nominal case, and can in theory be controlled by varying the relative inputs to the rotors. This configuration also minimizes the gyroscopic moment on the tilting mechanism, allowing the use of a smaller actuator.

Propeller Pitch

We assume according to our propeller model defined in Sect. 3.1.2 that rotor drag torques will be proportional to the thrust force. The propeller pitch is the major factor influencing the drag coefficient, and can be selected from a variety of standard pitch values. The drag coefficient influences both flight efficiency and performance, since higher drag requires more power from the motor, but also provides a higher control authority with its torque. For double rotor groups that we model as a single pure thrust generator, drag torques do not provide additional control authority. Higher pitch propellers still provide more aggressive flight, and are favoured under our assumption that rotor forces can be instantaneously achieved.

Rotor Speed Control

Each BLDC motor receives 3-phase power from an ESC, which determines the rotor speed. In our implementation of double rotor groups, the chosen ESC is the T-motor F45A which takes feed-forward PWM commands as input, and has a 45 A current capacity to accommodate the expected thrust transients. This choice is largely due to market availability, meaning an extensive product range and multiple suppliers, as well as a relatively low price.

Improvements on the standard PWM communication model have been available in the hobby community for some time. The OneShot protocol divides pulse widths by a factor of 8, significantly increasing the speed of communication from the controlling computer to the ESC. A more recent improvement is the DShot digital communication protocol, allowing fast transmission of high-resolution digital commands, without the need for ESC throttle range calibration. While the advantages of these protocols are desirable, the ESCs produced for this purpose are designed for a quadcopter, and the integration and cabling of these systems for a 12 rotor system is complex and daunting.

A recent introduction to the ESC market for MAV parts is the velocity ESC. Closing the loop on angular velocity at a high internal rate using rotor velocity feedback, these ESCs accept direct velocity commands, removing the need for PWM to rotor speed mapping and compensation for changes in supply voltage. Furthermore, these systems are controlled with a controller area network (CAN) interface, simplifying the system integration with a single bus connection for all ESCs and providing real-time telemetry data of motor speed, current, voltage, and temperature to the controlling computer. This additional feedback allows for health monitoring of the system. Since they are rather new to the hobby market, the product range, price and availability are the disadvantages that have prevented us from choosing this solution thus far.

Mechanical Integration

Integrating a double rotor group into the design requires rigidly mounting these components the system at the desired spacing of approximately 0.3 times the propeller diameter ($\approx$ 7 cm spacing for a 9 inch propeller), to reduce efficiency loss caused by interfering airflows.

Each motor is rigidly mounted to a small carbon fibre plate, offset with aluminium spacers. These plates are then attached to 16 mm carbon fibre arm tubes with two aluminium clamps. Figure 5.4 shows exploded CAD models to illustrate the rotor group assembly in detail. Propeller airflow generated by the rotors themselves cools the motors during flight. The resulting double rotor group mass is 0.25 kg, and inertia about the tilting axis is 3.5×10^{-4} kg m^2. Note that these values also include the rotating support arm and adapter described in Sect. 5.1.2. Damped mounting may be considered in the future to reduce high-frequency vibrations at the system

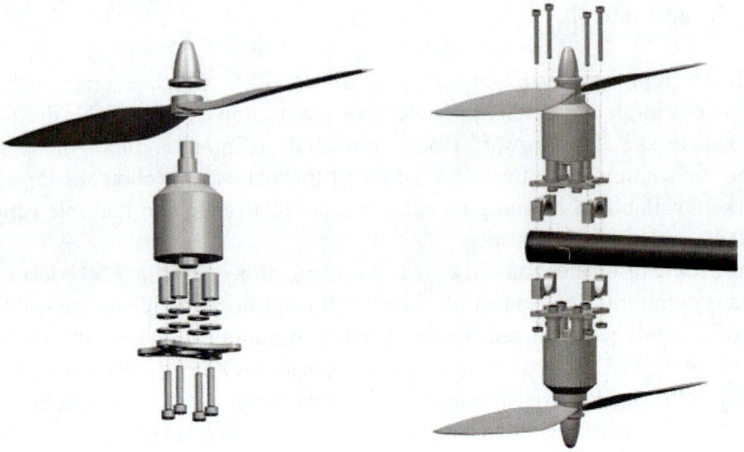

Fig. 5.4 Assembly of a double rotor group, where individual rotors are attached to a small carbon fibre plate with aluminium spacers (left), then rotor groups are clamped together on a carbon fibre supporting arm tube

core. Damper integration usually results in increased weight, and must be carefully designed to eliminate only high-frequency vibrations.

5.1.2 Tilt-Rotor Mechanics

The integration of a tilting mechanism can theoretically occur at any point along the $\boldsymbol{e}_x^{R_i}$ axis (rotational axis for the ith rotor group), which is collinear with the supporting arm. The mechanism consists of a tilting motor and transmission mechanics such as bearings and clutches. We consider implementing this mechanism at the rotor group location (as done in [2]) and at the system base.

The motor mass and transmission mechanics are non-negligible in weight, and integrating them at the rotor groups increases the overall system inertia, which subsequently reduces flight agility. In addition, with a prototype destined for research on novel controllers, we want the system to tolerate crashes to some extent, and to be easily repairable if these crashes are catastrophic. For these reasons, integrating the tilt-rotor mechanism at the base is s good solution to reduce overall system inertia, and keep most of the mechanics in a protected base core.

For serviceability, the output of the tilt mechanism interfaces to a carbon support tube of 16 mm outer diameter and 0.5 mm. An internal aluminium sleeve prevents crushing of the tube when clamped into the transmission adapter, and the connection assembly is shown in Fig. 5.5. Since rotor groups are clamped to the outside of the tube, in the event of a crash these can be easily remounted to a new support tube.

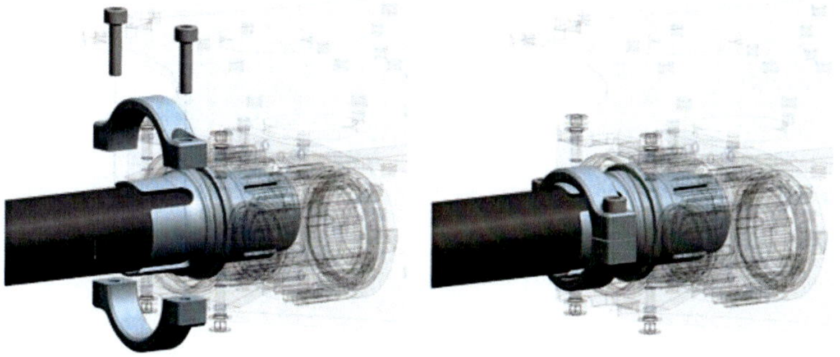

Fig. 5.5 Integration of the carbon fibre rotor group support arm in the tilting mechanism, with a designed break-point at the clamping edge of the carbon tube

When choosing a motor for the tilting mechanism, we narrow down a wide range of options by requiring that the motor

- takes commands and provides absolute feedback in position and velocity,
- has integrated motor control with no additional hardware requirements,
- has sufficient torque and speed specifications, and
- is relatively low in cost.

Smart servomotors almost exclusively fulfil these requirements, and as such we select our tilting actuators from this category. In an effort to maintain low weight with sufficient performance criteria, we select the XL430-W250 model which at a nominal 12 V supply has a no-load speed and current of 6.38 rad s^{-1} and 0.15 A respectively, and stall torque and current of 1.5 N m and 1.4 A respectively. In the worst case, the current draw for 6 motors is 8.4 A, but in nominal flight we expect it to be insignificant compared to the draw from rotor groups.

Tilt-Rotor Transmission and Cabling

Our complete tilting mechanism must be compact for integration into the system base with other required components, and cables that connect to the tilting rotor groups must be managed to allow for tilt movement without interfering with propellers. We choose to integrate ESCs at the base, such that small signal cables are protected within the system core, and 3 phase power for the BLDC motors run along the support arm. We develop two designs for implementation in the prototypes, which sit at two points on the design trade-off of complexity, cable management, and ease of repair. A summary of their relative advantages and drawbacks is shown in Table 5.2.

The direct transmission design shown in Fig. 5.6 consists of a main aluminium shaft that transmits rotary motion from the servo-motor to the rotor group support arm. Rotary motion is facilitated by two steel ball bearings in custom aluminium housings that connect to the surrounding core structure. For both direct and offset

Table 5.2 Design trade-offs for direct and offset tilt mechanism implementations

Direct transmission	Offset transmission
+ minimal backlash	– backlash in gearing
+ simple design	– higher complexity and weight
+ trivial arm replacement	– arm swap with cable routing
– cable wind-up one rotation	+ cable wind-up 5 rotations
– large external service loops	+ cables routed internally

implementations, we protect the servomotor from shock damage with a spider clutch, where a compliant 'spider' securely fits between adapters on the motor and output shaft, and transmits rotary motion with minimal backlash. This construction is clearly visible in Fig. 5.6a.

The offset transmission design shown in Fig. 5.7 consists of dual aluminium shafts for the motor input and tilt-arm output. These interface with nylon 1:1 gears, each glued onto their respective shaft with complementary grooves to prevent rotation. Gear alignment and rotary motion is provided by two bearings on each shaft which are aligned with an integrated bearing housing that connects to the surrounding core structure. Both direct and offset constructions make use of the servomotor's large surface area and attachment points to act as additional structural support for the system core.

Infinite Tilting

As a preliminary work, we have further explored a tilt mechanism design that allows for infinite tilting of the joint, freeing the system from constraints due to cable wind-up. While off-the-shelf slip rings capable on handling expected rotor currents are much too heavy for installation on an aerial platform of our size, this new concept involves the use of rotary brass bearing surfaces as lightweight slip rings for power transmission.

Implementing a slip-ring directly on the three-phase power cables that control the BLDC motors introduces voltage losses and resulting model error for rotor thrust control, and for a double rotor group would require 6 contacts. Instead, we choose to transmit the positive and negative battery contacts, placing the ESCs inside the tilting arm, as modelled in Fig. 5.8. Two rings work nicely for the mechanical function and secondary use of the slip-rings as load-supporting bearings.

This setup leaves us with signal cables that must transmit commands from the controller in the core to the tilting ESCs. Since these signals use low current, an off-the-shelf slip ring is well sized for this problem, and is integrated at the base of the tilting arm. Since cables still run through the arm axis, we can use a geared tilt-motor to allow cables to run through, or as in this prototype, select a hollow shaft motor for minimal complexity. Currently available hollow shaft motors are significantly

(a) Exploded component diagram of *direct* tilt-rotor transmission.

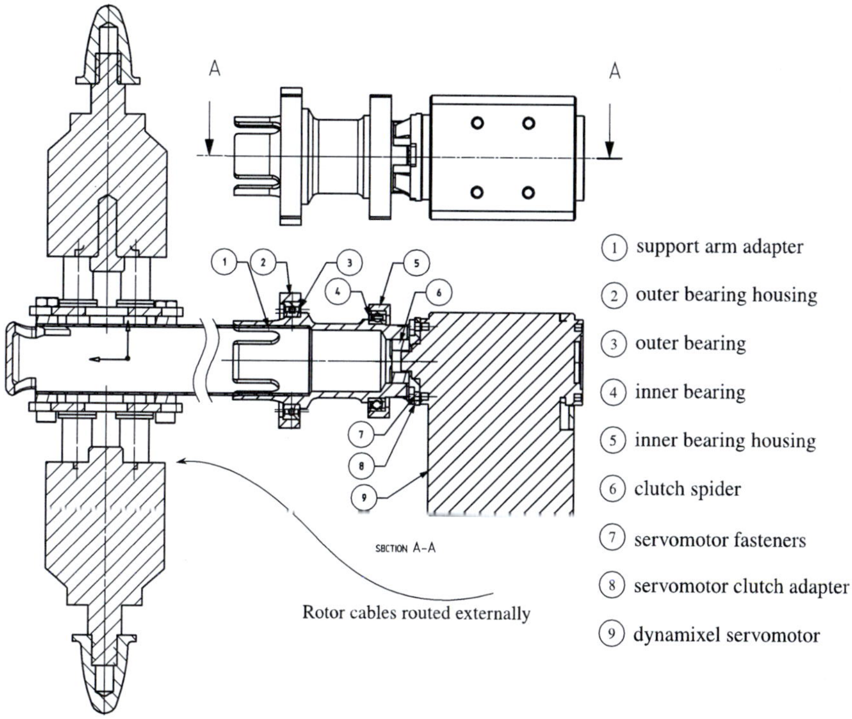

(b) A section-view drawing indicates major direct transmission components. The cable routing path is indicated, passing outside the transmission the rotor groups. As a result, the cables wind around the supporting arm as it turns.

Fig. 5.6 Direct tilt-rotor transmission

heavier than servomotors for the same performance, but will be an elegant solution if technology improves in this direction.

A physical prototype shown in Fig. 5.9 was constructed as a proof of concept, and preliminary thrust test results in Fig. 5.10 demonstrate the feasibility of combined rotary motion and power transmission. Electrical resistance across the power-transmitting bushing was calculated during bench tests for different grease and pre-load conditions, resulting in a resistance as low as $0.2\mathrm{m}\Omega$ under pre-load with silver conductive grease.

(a) Exploded component diagram of *offset* tilt-rotor geared transmission.

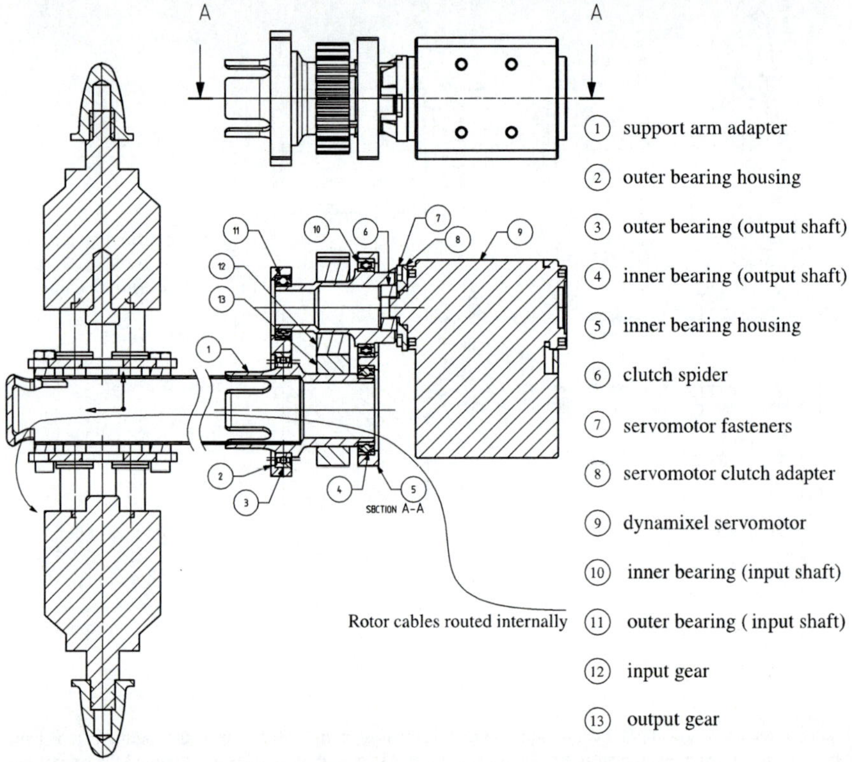

(b) A section-view drawing indicates major offset transmission components. The geared construction creates an offset between the motor and output shafts, allowing cables to pass through the support arm tube to the rotors as indicated. As a result, cables twist within the arm as it turns, allowing multiple rotations and shorter overall cable length than the direct transmission in Fig. 5.6.

Fig. 5.7 Offset tilt-rotor geared transmission

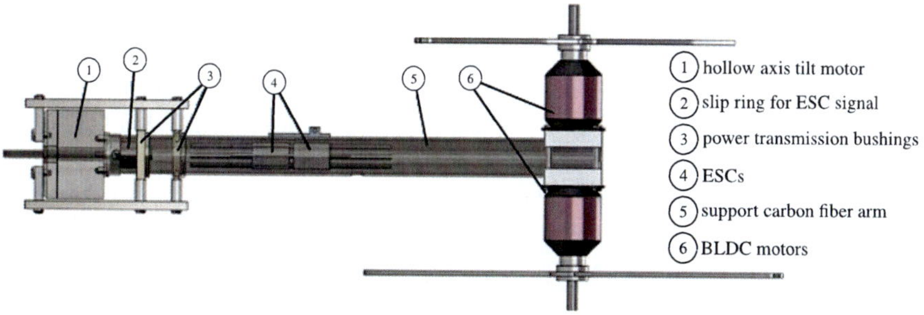

Fig. 5.8 A CAD model of an infinitely tilt-able rotor unit with a power-transmitting bearing. Positive and negative power are transmitted via independent brass bushings to ESCs inside a carbon fibre support arm

Fig. 5.9 Physical prototype of the complete double tilt-rotor assembly (left) and a close-up of the power transmitting bearings, isolated from the surrounding structure by nylon stand-offs

Fig. 5.10 Thrust bench test results of an infinitely tilt-able double rotor unit

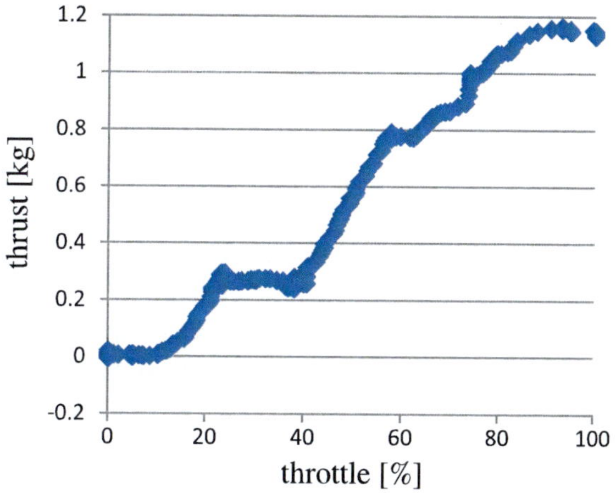

As this technology is at an early stage of development, extensive long-term testing should be performed before implementation on a flying robot. This concept also presents the danger of transmitting electricity through exposed electrical contacts while flying near humans, or in case of a crash. Further design work should develop a protective cover around the electrical surfaces, and electronic protection of the battery should be implemented to prevent a short circuit.

5.1.3 System Core

The core of the flying system contains the hardware, electronics, and power required for autonomous flight, contained within a compact structural casing. Here we consider the purpose of their integration and the overall core mechanical assembly.

Computation

Our system works with two stages of computation. An onboard computer (Intel NUCi7) handles autonomy, including planning, state estimation, and high-level control, and communicates with a base station to receive operator input and motion capture odometry estimates. Its high processing capacity also makes it suitable for image processing required for interpreting point clouds of local surfaces and implementing state estimation with visual-inertial odometry (VIO).

A flight micro-controller (Pixhawk mRo/4) handles low-level flight control, and implements safety features via a direct radio connection to an operator's Radio Control (RC). The hardware offers ports sufficient to control 12 rotors and 6 chained servomotors, but with limited processing power and memory, it is only suitable to run controllers with low processing requirements.

Power

The onboard power requirement is dominated by the rotors which work constantly against gravitational forces to maintain hover. Although we would ideally have batteries with the maximum possible capacity, their significant weight drives a tradeoff, and largely influences the design iteration process. We use '6S1P' (6$\times$ 3.7 V cells connected in series) LiPo batteries, for high power density, with high enough voltage to maintain reasonable current values throughout the system. Based on our selection of double rotor groups, we estimate an average power draw of 45 A and 60 A at hover in the most and least efficient configurations respectively.

For a 7000 mAh battery at 22.2 V and 25C maximum discharge rate (175 A), current draw in hover results in 7 to 9.3 min of flight time for a system with a full payload.

In earlier prototypes, two 3800 mAh batteries are installed and connected in parallel. The two batteries can be spaced apart on the system to make room for additional components at the center in a compact arrangement. Though this solution works in flight, mismatched age and condition of the two batteries can significantly reduce overall flight time, as well as battery health. For later prototypes, a single 7000 mAh battery is installed, which greatly simplifies battery management. Assuming that most of the task time is spent transporting the system to and from the operation site, the flight time is sufficient for demonstrations and preliminary research investigations. For long-term inspection tasks, a tethered power cable is an ideal, solution, transmitting the required power to the system at high voltage, which is converted at the base with an on-board transformer. While the transformer may weigh similar to a battery, flight time is not restricted. This solution is particularly suitable for tasks such as NDT concrete inspection, since tethering is also required for sensor signal and water cables.

Sensors

Additional sensors are installed in the core assembly to enable system autonomy. A high quality inertial measurement unit (IMU) (ADIS 16448) is synchronized with a small camera with high dynamic range using a VersaVIS board [3] for VIO in a Rovio framework [4]. A Pico Monstar TOF camera is mounted to the core structure facing the direction of desired interaction, to provide local surface information for planning and for distance measurements used for hybrid force-impedance control.

Core Assembly

The above-mentioned components are integrated into a compact core structure. Two carbon fibre base plates sandwich the six tilt-rotor groups, connecting to bearing housings and servomotors for a strong base that approximates a rigid body. Direct and offset tilt mechanisms take up different volumes inside the core, and result in slightly different component arrangements.

Here we describe a compact arrangement for offset tilt mechanisms, with exploded component diagrams shown in Fig. 5.11. The main computer mounts to the lower base plate with nylon stand-offs. Landing gear is also connected here, consisting of nylon adapters for 10 mm carbon fibre tubes, and foam hemispheres to absorb impact. The Pixhawk microcontroller and IMU are mounted on a damped plate, which is suspended from the top plate on nylon stand-offs. The IMU position is located at the geometric center of the platform to reduce the need for offset calibration. Dampers reduce the effect of high-frequency propeller vibrations on ADIS and Pixhawk internal IMU readings. The power board is mounted directly on the upper plate, minimizing the distance to the batteries. For the dual battery implementation, these are located on either side of the power board with approximately symmetric inertia on this plane, while the single battery solution places it directly over the power

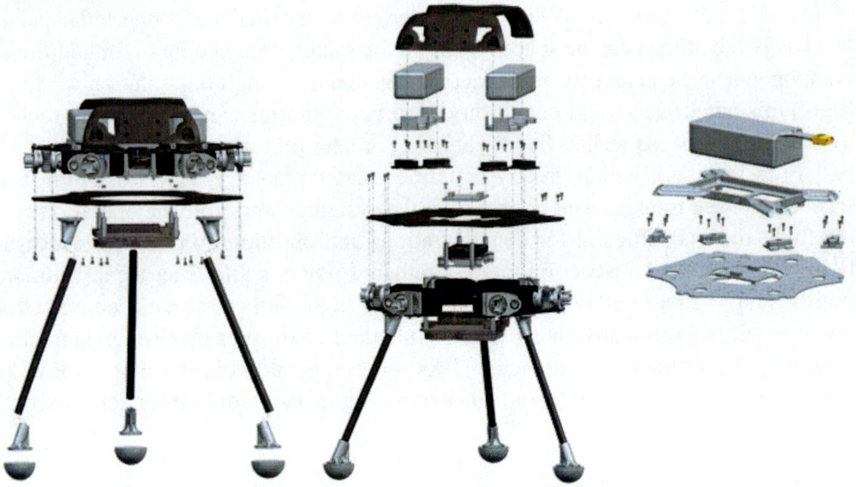

Fig. 5.11 Exploded component diagrams of the core assembly with offset tilt mechanisms. **Left:** Lower base plate with mounted computer and landing gear. **Center:** Upper base plate with suspended and damped flight controller and IMU unit, power board flanked by dual batteries, and protective cage with sensor mounts connected via adapters. **Right:** Single battery mounting configuration with clearance over the power board

board. A protective cover on the system mounts to adapters on the top plate, such that it can be removed without affecting the rest of the system. A similar cover can be attached below the system to protect the computer and provide mounting points for sensors.

5.1.4 Fixed Manipulator

The simplest solution to transmit forces from the base to an end-effector point is a static manipulator arm that is rigidly mounted to the base.

Primary design requirements for this arm include high stiffness, and length sufficient to distance the interaction point from spinning propellers. Since no state feedback of the end-effector point is available, any flexibility of the arm will directly limit our ability to track the end-effector point. Furthermore, the dynamics of a flexible arm could add error and compromise flight stability, since we do not consider flexibility or vibrations in the system model. The secondary design requirement is to minimize total weight. Since rotational inertia increases with the square of the distance of the added mass, weight of the arm has a significant effect. The overall mass also affects the system's endurance, reducing the flight time for a given battery size.

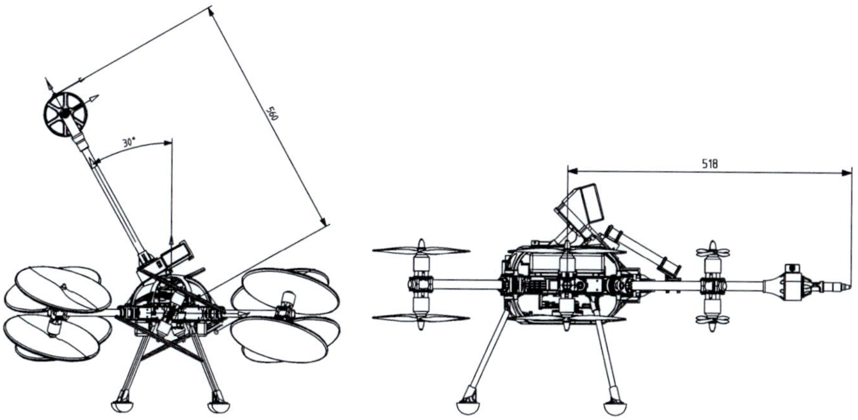

(a) Angled manipulator design. (b) Forward manipulator design.

Fig. 5.12 Fixed manipulator designs

An excellent material candidate for low weight and high stiffness is carbon fibre, and we choose a strong tube structure for the manipulator arm. Custom 3D printed polyamide adapters form connections with the base and end-effector. Figure 5.12 shows two realizations of a fixed manipulator arm design.

Integrated Force Sensing

To enable direct force control, at the end-effector point, we integrate a Rokubi 6-axis force-torque sensor just before the end-effector attachment. At 120g, the sensor weight is small enough that stable flight can be maintained. An alternate solution would be to place the sensor at the base of the interaction arm. In this solution, however, lateral forces at the end-effector create large torques on the sensor that can saturate and skew linear force readings. In addition, the arm's dynamics should be compensated to extract only the contact forces, requiring an additional modelling step.

End Effectors

For aerial physical interaction experiments, we use a series of different end-effectors, installed depending on the task.

Figure 5.13 shows prototypes of the following four end-effectors:

- **Foam ball:** for point interaction. High friction with surfaces for point force application.

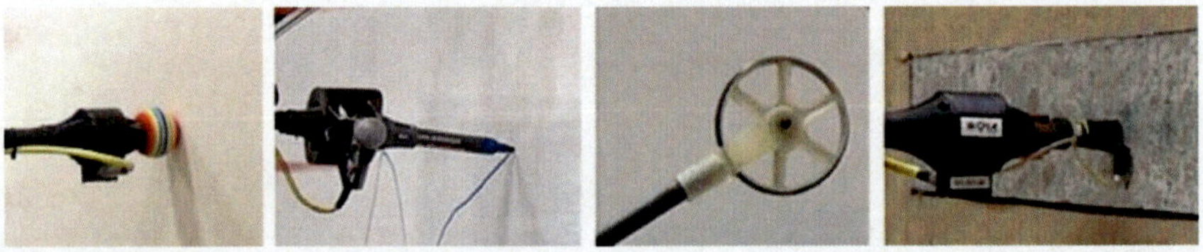

Fig. 5.13 End effectors, from left to right: foam ball, whiteboard marker, wheel, and CSE inspection sensor. All except for the wheel are mounted on a force-sensing adapter

- **Whiteboard marker:** for aerial writing applications. Low friction between the marker and whiteboard surface allows for push-and-slide contact tasks.
- **Wheel:** for rolling-in-contact along arbitrary surfaces with low friction in the wheel's direction of travel.
- **CSE sensor** for NDT point inspection of steel-reinforced concrete, requiring tethered signal and water cables between the flying robot and a base station.

5.1.5 Delta Parallel Manipulator

Following the optimization of geometric parameters for a delta manipulator presented in Sect. 3.4.2, we develop a complete mechanical design for the parallel structure. The manipulator prototype is designed for high rigidity, minimal joint friction, and precise geometric realization.

The distal link parallelogram is constructed using 6 rotational ball bearing joints as shown in Fig. 5.14. A first iteration of the structure used plastic spherical joints at each corner of the parallelogram (Fig. 5.14a). The final design (Fig. 5.14b) uses single-axis ball bearings, which significantly reduce joint friction, and increase the angular joint range from 66°–112° where components of the first iteration would deform elastically at the joint limits. A wider parallelogram improves model fidelity

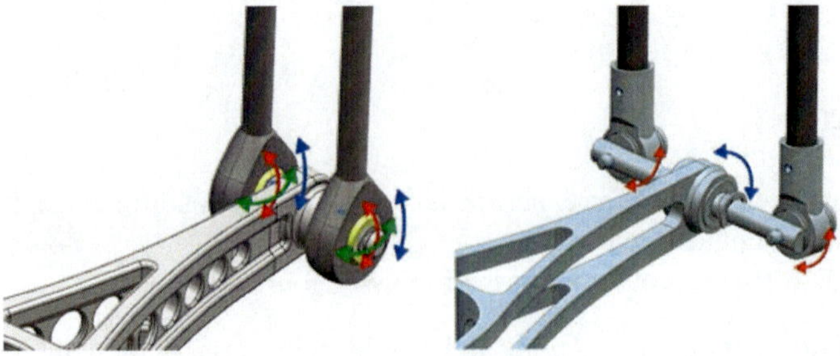

(a) Two universal joints. (b) Three rotary bearing joints.

Fig. 5.14 Mechanical design iterations of distal link parallelogram

Table 5.3 Delta manipulator system and component masses

Component	Qty	Unit mass [kg]	Mass [kg]
Omni MAV	1	3.3	**3.3**
Battery	1	0.8	**0.8**
Delta base	1	0.143	**0.143**
Servo motor	3	0.1	**0.3**
Proximal link	3	0.039	**0.117**
Elbow/wrist	6	0.004	**0.024**
Distal link	6	0.018	**0.108**
Tool plate	1	0.023	**0.023**
Moving Total			**0.272**
System Total			**4.82**

at large lateral angles. The distal links are hollow carbon rods, which are fixated with a length jig to reduce model error.

A detailed summary of components, along with their quantities and masses, is presented in Table 5.3, and a drawing of the system and image of the final prototype are shown in Fig. 5.15. The complete manipulator assembly including the mounting base weighs 0.72 kg, of which only 0.28 kg is moving relative to the base. Mounted on a 4.1 kg flying robot, the proportion of moving to total mass is less than 6%, confirming the inertial advantage of the parallel manipulator over a serial design. Dynamixel M430-W210 motors are selected for their sufficient torque capability and integrated position and velocity feedback. Since this manipulator is not yet used for contact with the environment, a foam ball end-effector is installed for a better visual of the end-effector reference tracking.

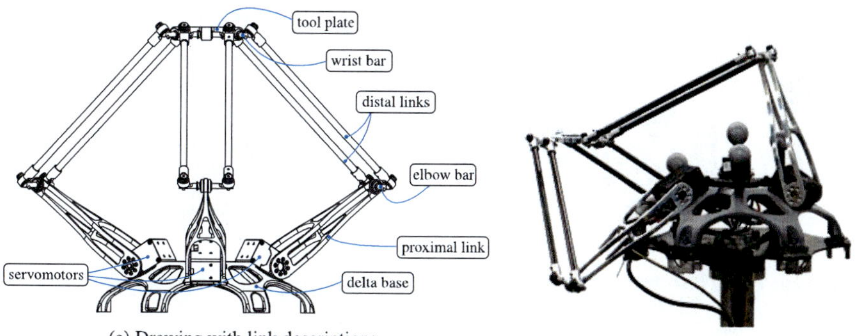

(a) Drawing with link descriptions.

(b) Final prototype Manip$^{\Delta}$.

Fig. 5.15 Delta parallel manipulator final design

5.2 Electronics

Complete integration requires connecting power and communication across the system. In this section, we summarize the power distribution and communication of the system at a high level.

5.2.1 *Power Distribution*

A LiPo battery (or battery pair) provides power to all components on board the system. Arriving at the power distribution board at 22.2 V, voltage converters step the voltage level as required for different components, as shown in Fig. 5.16.

The vast majority of battery power goes directly to 12 ESCs without transformation, each of which provides 3 phase power to a BLDC motor. Transformation to 12 V is required for servomotors that drive the tilt-rotor joints and the delta manipulator. Power is chained for each motor group, reducing the overall required cabling. The flight controller requires 5 V and the onboard computer 18 V.

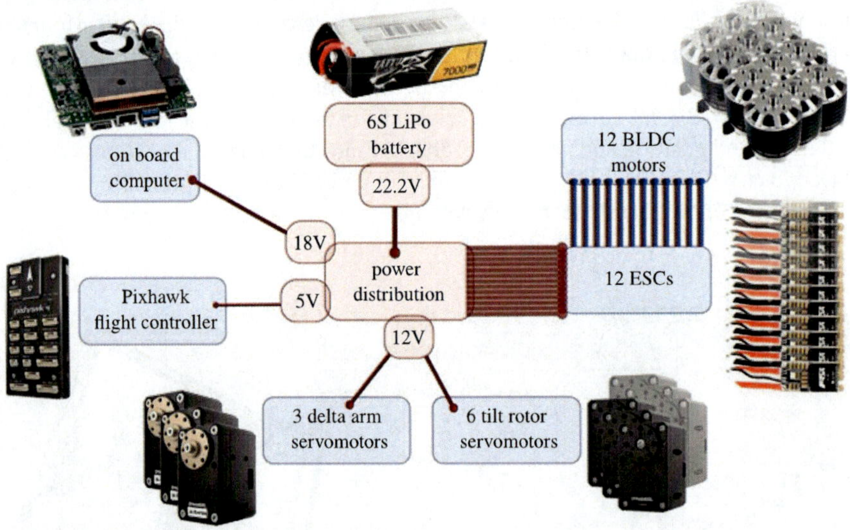

Fig. 5.16 Power distribution summary for essential flight components

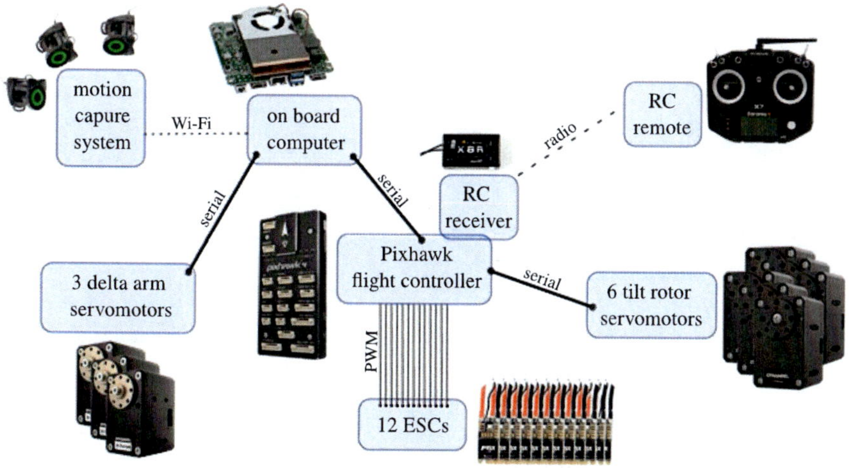

Fig. 5.17 System communication connectivity summary for essential flight components

5.2.2 *Communication*

Critical core components for autonomous flight described in Sect. 5.1.3, are connected for communication as shown in Fig. 5.17.

The onboard computer receives motion capture (mocap) odometry estimates via WIFI, and performs state estimation, scheduling of pre-defined trajectories, and high level (actuation wrench) control. When a delta manipulator is integrated, the controller also directly commands the delta servomotors over a serial connection. The computer is also connected by a serial bus to the flight controller, which performs lower-level (actuator allocation) control. PWM commands are then sent directly to the 12 ESCs, and position commands are sent to the 6 servomotors over a serial bus.

The Dynamixel servomotors are chained in power and communication, which greatly simplifies overall cabling, but if damage occurs at any point along the cable, many tilt motors can have unexpected performance. We have addressed this issue with a star-configuration cable, ensuring that every motor has two communication paths bask to the flight controller. Cabling of the ESCs is individual, resulting in a difficult cable routing problem in the system's interior. Moving to a chained ESC protocol such as UAV-CAN would not only help cabling, but would also provide valuable feedback to the flight controller.

The Pixhawk flight controller also has an RC receiver which communicates over radio with an RC remote, operated by a safety pilot. This communication element is critical in terms of safety, enabling instantaneous shut-off of all actuators commanded by the flight controller. In case of unexpected flight, the system can be stopped quickly regardless of the state of the computer.

(a) Proto^{V1}. (b) Proto^{V2}. (c) Proto^{Δ}.

Fig. 5.18 Images of main prototypes with integrated fixed, force sensing and active manipulators

Software Framework

Software on the on-board computer is written in C++ and python, and integrated in a Robot Operating System (ROS) framework. Different nodes, such as a mocap receiving, state estimation, and wrench control, communicate using ROS topics. A designated node also handles the conversion of control commands to the flight controller, a Nutt-X-based system with uORB communication topics.

5.3 Overview of Prototypes

Over the course of the doctoral studies summarized here, several prototype systems were developed and tested, combining the mechanical and electrical design details discussed in this chapter.

Figure 5.18 depicts the three main prototype iterations that are used for experimental evaluation of the prototypes and controllers, including various integrated manipulators. The final prototypes are referred to in Chap. 6 as follows:

- $\text{Proto}^{V1}$is the original tilt-rotor system prototype, distinct for its older Pixhawk mRo flight controller and its use of direct tilt transmission mechanisms.
- $\text{Proto}^{V2}$is an upgraded system with offset tilt transmission for internal rotor cable routing, and the newer Pixhawk 4 flight controller.
- $\text{Proto}^{\Delta}$is the same as Proto^{V2}, now using a single 7000m A h 6S LiPo battery instead of 2× 3800mAh batteries, and a delta parallel manipulator.
- $\text{Manip}^{\Delta}$is the independent delta parallel manipulator as shown in Fig. 5.15b, connected to and controlled by an external computer.

with main components for the tilt-rotor MAVs listed in table 5.4.

General inertial properties for the tilt-rotor prototype without additional sensors are listed in Table 5.5.

Table 5.4 Major system components, and differences between prototypes. A left arrow indicates component is the same as for Proto^{V1}

Component	Qty	Proto^{V1}	Proto^{V2}	Proto$^{\Delta}$
Onboard computer	1	Intel NUC i7	←	←
Flight controller	1	Pixhawk mRo	**Pixhawk 4**	**Pixhawk 4**
BLDC rotor	12	KDE 2315XF-885	←	←
Propeller	12	Gemfan 9 × 4.7	←	←
ESC	12	T-motor F45A	←	←
Tilt motor	6	Dynamixel XL430-W250	←	←
Tilt mechanism	6	direct	**offset**	**offset**
IMU	1	ADIS16448	←	←
Battery	1	2× 3800mAh 6S LiPo	←	**7000mAh 6S LiPo**

Table 5.5 Main system parameters. *Primary components of inertia are obtained from CAD model

Parameter	Value	Units
Rotor groups	6	
Arm length, l	0.3	[m]
Total mass	4.27	[kg]
Inertia*	{0.086, 0.088, 0.16}	[kgm^2]
Diameter	0.83	[m]
$f_{R_i\,max}$	11	[N]
c_f	7.1e−6	[Ns^2/rad^2]
ω_{max}	1250	[rad/s]

References

1. LLC, K.D.: Kde direct (2021). https://www.kdedirect.com/
2. Kamel, M., Verling, S., Elkhatib, O., Sprecher, C., Wulkop, P., Taylor, Z., Siegwart, R., Gilitschenski, I.: The voliro omniorientational hexacopter: an agile and maneuverable Tiltable-Rotor aerial vehicle. IEEE Robot. Autom. Mag. **25**(4), 34–44 (2018)
3. Tschopp, F., Riner, M., Fehr, M., Bernreiter, L., Furrer, F., Novkovic, T., Pfrunder, A., Cadena, C., Siegwart, R., Nieto, J.: Versavis-an open versatile multi-camera visual-inertial sensor suite. Sensors **20**(5), 1439 (2020)
4. Bloesch, M., Burri, M., Omari, S., Hutter, M., Siegwart, R.: Iterated extended kalman filter based visual-inertial odometry using direct photometric feedback. Int. J. Robot. Res. **36**(10), 1053–1072 (2017)

Chapter 6
Experimental Results

In order to validate the modelling, control and design work presented in the previous chapters, and to evaluate the tilt-rotor aerial robot as an effective omnidirectional aerial manipulation platform according to our problem definition, we perform flight experiments that touch on many of these presented concepts.

Our experiments are divided into the three major categories of free flight with an omnidirectional tilt-rotor flying base, aerial interaction of the system with a rigidly mounted manipulator, and macro-micro end-effector tracking with a parallel manipulator mounted to the omnidirectional base. Major objectives to be evaluated are listed in Table 6.1, and corresponding video material is listed in Table 6.2.

Presentation of Results

Position and attitude errors throughout the section are taken from data collected over the complete duration of the trajectory, and accumulated for all successful tests with a given set of parameters where multiple trials have been performed. Since we are interested in exploiting the omnidirectional capabilities of the system and thus track all 6 axes of a trajectory, errors are given separately in all axes where applicable. The norm of the position errors is the Euclidean error, and the angle error for rotation is computed as the norm of the rotation vector derived from $\boldsymbol{R}_{WB}^{*}\boldsymbol{R}_{WB}^{\top}$, which is equivalent to the angle from an axis-angle representation. Tables of error statistics present the mean, standard deviation and RMSE for each value. We use the mean and standard deviation to compare position error norm and total angle between trials, and the RMSE to compare error magnitudes by axis within a given trial.

Position in the plots is shown in meters, and discussed in the text in centimeters or millimeters for more intuitive interpretation. Attitude is usually expressed in radians, except in certain cases where a description in degrees better facilitates interpretation. Both units of attitude are used to express yaw-pitch-roll Euler angles of the system, and attitude error is presented as the difference from set point values. Error values

K. Bodie et al., *Omnidirectional Tilt-Rotor Flying Robots for Aerial Physical Interaction*, Springer Tracts in Advanced Robotics 157,
https://doi.org/10.1007/978-3-031-45497-4_6

Table 6.1 Objectives of experimental evaluation by category

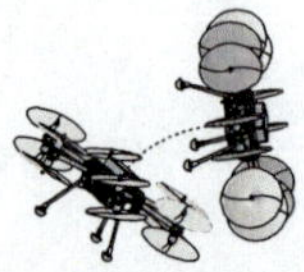 Omnidirectional Flying Base (section 6.2)	• Omnidirectional trajectory tracking • Robustness to singular configurations • Unwinding of tilt-arm cables • Flight efficiency evaluation
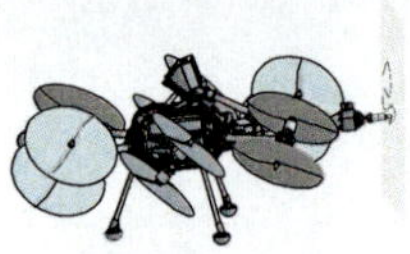 Aerial Physical Interaction (section 6.3)	• Evaluate external contact wrench estimate • Disturbance rejection with axis-selective behaviour • Direct interaction force control • Push-and-slide interaction with a planar surface • Validate system viability for NDT contact inspection • Interaction with inclined and non-planar surfaces
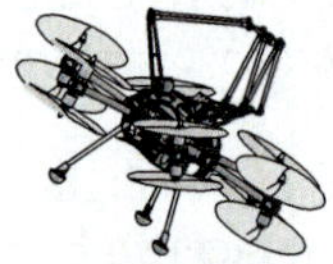 Macro-Micro Manipulation (section 6.4)	• Validation of parallel manipulator dynamics • Disturbance rejection with macro-micro system • Evaluate performance for fast end-effector tracking

Table 6.2 Video links for flight experiments

Experiments	Section	Video link
Omnidirectional flying base	6.2	https://youtu.be/FcxVmAa3HYo [1] https://youtu.be/FcxVmAa3HYo [2]
Aerial physical interaction	6.3	https://youtu.be/-RCQmaKvsL0 [3] https://youtu.be/7Nvlki1xo-c [4]
Macro-micro manipulation	6.4	https://youtu.be/mAdFYjHcBrk [5]

are evaluated in violin plots which represent the probability density function for the error data. An overlaid box plot shows the median, the upper and lower quartile. Probability density functions are cut off at the extreme measured points.

6.1 Experimental Setup

To evaluate the proposed aerial systems and control strategies, our experimental setup consists of a test environment and a prototype system (described in Sect. 5.3).

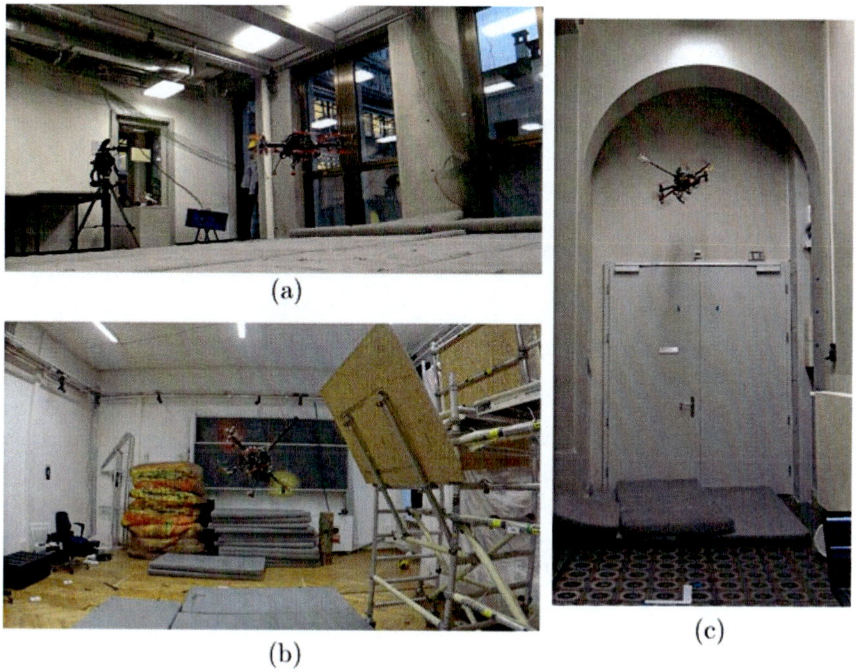

Fig. 6.1 Images of **a** small flight arena (small arena), **b** large flight arena (large arena), and **c** stairwell archway (archway) test environments

We further discuss the state estimation implementations and trajectory generation methods used throughout these experiments.

6.1.1 Test Environment

Several test environments have been used for experimental flights, both indoors and outdoors, with and without external state estimation systems. Since experimental flights often involve testing new and untuned controllers, additional considerations are essential for the safety of the operators, and to reduce damage to the flying robot. The test environments referred to in the following experiments are shown in Fig. 6.1 and are described in detail below.

Small Flight Arena

A small flight arena with an area of approximately 4 m × 4 m and height of around 3 m is enclosed by floor-to-ceiling safety netting on three sides. A wall on one side of the arena can be used for mounting a whiteboard or other props for interaction. A Vicon motion tracking installation consists of eight cameras positioned around the room. A safety tether line (aramid braided cord, 3 mm diameter) is routed through carabiners on the ceiling to catch the system, preventing damage in case of system failure. A set of foam mattresses are available to cover the floor to reduce damage to the system and the environment in the event of a crash. The relatively small space results in non-negligible airflow circulation from propeller down-wash, which has an effect on performance metrics that is difficult to quantify.

Large Flight Arena

A large flight arena in a room approximately 8 m × 9 m and 4 m high. A Vicon motion tracking installation consists of 17 cameras positioned around the room. A safety tether is installed similarly to that of the small flight arena, and a set of mattresses is available to cover the floor. Several scaffolding structures are present to act as walls that can be arbitrarily positioned in the room. The large space generally has less airflow disturbance than the small flight arena.

Stairwell Archway

In order to simulate an overhead non-planar infrastructure inspection scenario, we select the ceiling vault of a staircase landing as a test environment. The maximum altitude of the arch surface is approximately 4 m above the ground. No external state estimation or safety tether is available in this scenario, so state estimation relies on VIO estimates.

6.1.2 State Estimation

All state estimation solutions used for flight experiments use a modular framework for MSF based EKF[1] to combine observations of the system's pose with proprioceptive acceleration and angular velocity measurements from an on-board IMU. We consider different sources of pose information for the state estimator including mocap, VIO and TST estimates. An overview of the specific implementations is shown in Table 6.3.

On board IMU sensor data are sampled at 250 Hz, and fed into the MSF state estimation module. Several different pose estimation strategies can be fused in, each with its benefits and drawbacks.

Pose measurements from a mocap installation (in our case, a Vicon system) are theoretically capable of sub-millimeter precision and accuracy, which is sufficient to

[1] State estimation source code is available at https://github.com/ethz-asl/ethzasl_msf.

Table 6.3 State estimation methods used in experiments

State estimator	Abbrev.	Characteristics
Fused mocap and IMU	Vicon-MSF	+ High precision and accuracy + High rate pose estimates − Large external infrastructure and calibration
Fused VIO and IMU	VIO-MSF	+ Completely on-board state estimation − Drift in position and yaw over time − Lower rate pose estimates
Fused TST and IMU	Leica-MSF	+ High precision and accuracy − External setup and calibration required − Line of sight to on-board marker

be considered ground truth data. Most experiments will be performed with this state estimation system to evaluate the nominal controller behaviour. The actual position and accuracy of the pose measurements are dependent on the system calibration, and the quality and arrangement of small reflective spheres in an asymmetric constellation that is mounted on the robot. Depending on the experiment, some of the markers may be occluded from the external cameras' line of sight, which can degrade the tracking performance. In real flight experiments of new controllers, periodic crashes can cause a slight change in marker position, affecting future state estimation performance. The major disadvantage of external camera motion tracking systems is the large external infrastructure required, and its calibration. Since line-of-sight is essential, experiments are limited to open spaces with a minimum of objects that may cause occlusion. For our experiments, this pose data is streamed at 100 Hz.

Where an external camera set-up is prohibitively difficult to install, pose sensing can be achieved with purely on-board sensing using VIO. The VI sensors used in our implementation contain a dedicated IMU and camera which are triggered simultaneously to collect time-aligned inertia and visual data. We use the Rovio[2] [6] framework to generate odometry estimates, which are computed at 20–40 Hz, and fed into the MSF state estimator. While the use of VIO relieves the need for external infrastructure and line of sight, its drawbacks include lower frequency pose estimates and gradual drift in position and yaw due to lack of an absolute reference. The algorithm further relies on a stationary environment for visual landmark tracking, and can diverge in rare cases when features are repetitive or cannot be reliably found.

Outdoor flight in open spaces can benefit from the integration of GNSS position estimates, with dual sensors to also provide an estimate of the heading. With high-quality sensors, and the addition of a RTK ground station, these measurements can be highly accurate at a high rate. In our intended scenario of infrastructure inspection, however, these methods suffer from multi-path effects that can cause pose estimates to jump in the areas where we require the highest precision. A solution to this is the use of an external observation point to provide a high-precision location and eliminate position drift. A TST can provide this functionality when a retro-reflective

[2] Rovio source code is available at https://github.com/ethz-asl/rovio.

marker is placed on the flying robot (in our case a prism for directional indifference), and an observation point is carefully chosen to ensure that line of sight is never occluded. This setup provides a high-rate and highly precise position estimate, but requires calibration, and line-of-sight tracking may not be recoverable if interrupted at any point in the operation.

Further state estimation options that would aid performance in realistic scenarios are localization within a map or SLAM, depending on our knowledge of the environment. Local planning based on sensor information is another solution, as long as a task can be interpreted in a local frame, such as the use of a TOF camera that estimates distance and orientation with respect to the contact surface.

6.1.3 Trajectory Generation

Trajectory planning is a critical element that influences the performance of an autonomous aerial system, and although relevant planning methods are not within the scope of this book, their implementation for experiments is briefly discussed in this section. Trajectories used in known environments in this work are computed based on smooth minimum-snap polynomial functions [7], with adaptations for 6D trajectories. No feasibility guarantees are considered, instead the trajectories are carefully designed to avoid collision, respect dynamic limitations of the platform, and position the system appropriately relative to a known structure for interaction.

Dynamically consistent trajectories are generated on the on-board computer from a series of pre-defined waypoints and stopping conditions. These trajectories are discretized at the control rate into an array of trajectory points. The controller iterates through the points sequentially. After trajectory completion, the last point is held, and velocity and acceleration references are set to zero.

Local Path Planning

For many contact interaction scenarios, a reliable outdoor-compatible localization system such as GNSS is not feasible, and a stationary reference such as a TST may lose line of sight. Without a global reference when interacting with a known structure, or in the general case of interacting with a structure where the geometry is unknown, a local surface observation is essential. In these situations, we use an onboard TOF camera to detect the local surface's distance and orientation, for semi-autonomous planning. A visual description of the local path planning method is shown in Fig. 6.2.

The system is first maneuvered to a pose facing the desired interaction surface, and a target contact point is chosen as the intersection between the $\boldsymbol{e}_z^E$ axis and the observed surface.

A small area of the surface around the target point must lie within the camera's field of view, and the observed 3D points within this area are used to compute the local surface normal and tangent plane. The contact set point is then transformed from

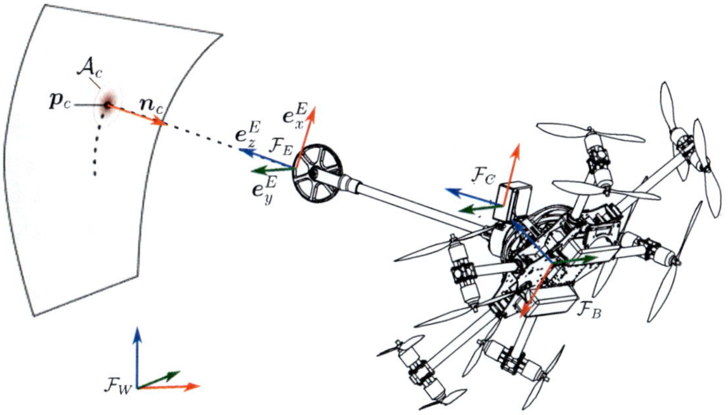

Fig. 6.2 Path planning on locally observed surface: p_c is the contact point defined by the intersection of the e_z^E axis and the observed surface, and n_c is the estimated surface normal at the contact point. $\mathcal{A}_c$ is the set of 3D points used to estimate the contact point location and normal. Dotted black lines indicate trajectories for approach and surface traversal based on these waypoints

the camera to the end-effector frame. If using a direct force controller, the set-point remains exactly on the observed surface, and a supplementary force target is specified in the direction of the surface. For interaction with an impedance controller, the target set point is placed a small distance behind the estimated contact point to generate a prescribed contact force, with its distance depending on the choice of impedance parameters. The target orientation of the MAV is chosen such that the e_z^E axis aligns with the estimated surface normal. Rotation about the e_z^E-axis is constrained by aligning the e_y^B axis parallel to the gravitational plane. The final axis fully defines the rotation as $e_x^E = e_y^E \times e_z^E$. A smooth trajectory is generated as previously described such that the end-effector is driven to the estimated target pose.

Once in contact, translation along the surface is achieved by moving the set point along the local tangent plane in a pre-defined direction. Constant observation of the local surface is required to update the tangent plane for a non-planar surface, or to correct drift if no global position observation is available. The implementation used for experiments updates the surface estimate and locally generated trajectory at a rate of 5 Hz.

Actuated Manipulator Trajectories

In experiments evaluating the macro-micro aerial system with a redundant parallel manipulator, separate trajectories are generated for base and end-effector motion, and supplied to their independent controllers. For the case of free flight and motion within the base's dynamic capabilities, a smooth 6 DOF trajectory is first computed for the task-relevant frame, in this case the end-effector:

$$\mathcal{T}_E = \{ {}^W\boldsymbol{p}_{E^*}(t), \boldsymbol{R}_{WE^*}(t), {}^W\boldsymbol{v}_{E^*}(t), {}^W\boldsymbol{\omega}_{E^*}(t), {}^W\dot{\boldsymbol{v}}_{E^*}(t), {}^W\dot{\boldsymbol{\omega}}_{E^*}(t)\} \quad \forall t \in [t_0, t_f]. \tag{6.1}$$

Given this trajectory, a corresponding smooth trajectory for the base is computed using a static transform, $\boldsymbol{T}_{B\bar{E}} = \{ {}^B\boldsymbol{p}_{\bar{E}}, \boldsymbol{R}_{B\bar{E}}\}$, that relates the base frame, $\mathcal{F}_B$, to the nominal end-effector pose, $\mathcal{F}_{\bar{E}}$, as described in Sect. 3.4.2. This transformation places the target end-effector position at the center of its dynamic workspace, for the ability to compensate disturbances effectively in all directions.

In cases where fast end-effector trajectories are desired which exceed the dynamic capabilities of the base, the reachable workspace of the manipulator is positioned such that these fast motions are feasible. Meanwhile, the base trajectory remains static or moves slowly to reposition the end-effector workspace.

6.2 Omnidirectional Flying Base

The first series of experiments evaluate the model, control and design of the aerial platform as an omnidirectional flying base. We evaluate the tilt-rotor prototype system Proto^{V1} in flight with a six DOF inverse dynamic PID controller (described in Sect. 4.1.2), with basic and differential actuator allocation (described in Sects. 4.1.3, 4.1.4). We perform the following series of experiments:

- Section 6.2.1 evaluates the tilt-rotor omnidirectional flying robot's ability to track **omnidirectional trajectories** with the baseline six DOF inverse dynamic PID controller. We evaluate the system's ability to perform isolated translational and rotational motion, as well as combined motion trajectories.
- Section 6.2.2 involves experiments that bring the system into **singular configurations**, and cases where tilt-arm **cable wind-up** becomes a concern. We evaluate the controller's ability to handle these conditions with two different actuator allocation approaches. We further evaluate the incorporation of winding awareness in the controller, to avoid a situation where an arbitrary omnidirectional trajectory causes the cables to wind up to the point where the rotation is blocked or the system is damaged.
- Section 6.2.3 evaluates data taken from some of the previous experiments to compare the system's measured **efficiency** to its expected efficiency based on the system model. Efficiency loss due to unwinding is presented for the case of differential actuator allocation.

The experimental setup for this series of test flights is described in Table 6.4, where Sects. 6.2.1, 6.2.2 each evaluate system properties for Proto^{V1} with PID-instantaneous control (a combination of PID wrench generation described in Sect. 4.1.2 and instantaneous allocation described in Sect. 4.1.3) and with PID-differential control (a combination of PID wrench generation and differential allocation described in Sect. 4.1.4). All experiments take place in the large flying arena

Table 6.4 Experimental setup for omnidirectional flying base evaluation

Experiment	Prototype	Controller	Test env.	State est.
Section 6.2.1A Section 6.2.2A	Proto^{V1}	PID-instantaneous	large arena	Vicon-MSF
Section 6.2.1B Section 6.2.2B	Proto^{V1}	PID-differential	large arena	Vicon-MSF

Table 6.5 Controller parameters for experimental flights

parameter	value
k_p	5
k_i	0 / 0.3
k_d	1.0
k_{p_R}	3.5
k_{i_R}	0 / 0.3
k_{d_R}	0.8

(a) PID control.

parameter	value	unit
φ_0	5	[°]
φ_d	15	[°]
φ_t	10	[°]
c_t	10	[°]
ω_u	8	[rad s^{-1}]

(b) Singularity handling.

parameter	value
k_α	1000
v_α	1
v_ω	250

(c) Differential allocation.

with Vicon-MSF state estimation. A safety tether is connected loosely to the robot, and minimally affects results.

Controller Parameters

Controller parameters and gains used in experimental tests are listed in Table 6.5, where we consider the parameters for a cascaded control structure including a PID wrench controller combined with an instantaneous or differential actuator allocation approach. Also included here are singularity-handling parameters that are described in Sect. 4.1.3. The weighting matrix $\boldsymbol{W}$ for the differential allocation experiments is blockdiag$(\mathbb{1}_{12}, k_\alpha \cdot \mathbb{1}_6)$. Integral gains are set to zero for experiments with instantaneous allocation, and to 0.3 for experiments with differential allocation, unless otherwise noted.

6.2.1 Omnidirectional Trajectory Tracking

Through two sets of experiments, we evaluate the tilt-rotor prototype system with a six DOF inverse dynamic PID controller for omnidirectional base trajectories.

Experiments in group A use an instantaneous actuator allocation scheme, performing trajectories that **isolate translational and rotational motion**, in order to demonstrate the ability of the chosen morphology and controller to separate translational and rotational system dynamics and motion tracking.

Experiments in group B use a differential actuator allocation scheme, performing trajectories that **combine translational and rotational motion**, to demonstrate the system performance for arbitrary six DOF motion, and verifying the nominal performance of a differential allocation approach away from kinematic singularities and unwinding conditions.

A: Independent Translation and Omnidirectional Rotation

To validate the controller's ability to isolate translational and rotational motion, and to evaluate its overall trajectory tracking performance in these cases, the following trajectories were performed:

1. Pure translational motion in $\boldsymbol{e}_x^W$ and $\boldsymbol{e}_y^W$ with constant flat hover orientation, over 17 s.
2. Pure rotational motion in roll, pitch and yaw simultaneously up to 45°, a duration of 14 s.
3. Pure rotational motion in a 360° flip about the $\boldsymbol{e}_y^B$ axis and back, a duration of 10 s per complete rotation.

Tracking results for trajectories 1 and 2 are shown in Fig. 6.3, where the plots on the left show the system's ability to perform translation while keeping a constant attitude, and the plots on the right show 3D rotation while tracking a constant position. A constant offset in the z position is due to the lack of an integral term, and corresponds to a model error related to the system, mass, thrust coefficient, or voltage compensation term. Despite this, we demonstrate the system's ability to effectively decouple its translational and rotational dynamics. In the case of the rotation trajectory, the world z position drops as higher angle deviations are commanded from flat hover. The absence of an integral term allows us to see this effect more clearly, demonstrating a model error for more extreme tilted configurations. Unmodeled effects such as

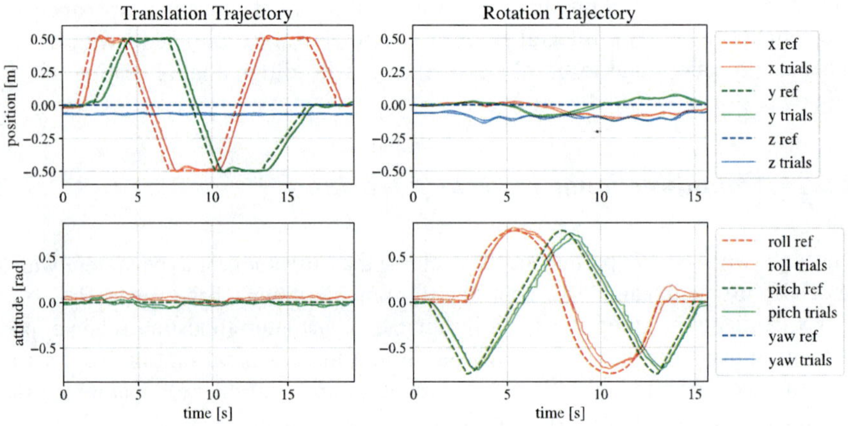

Fig. 6.3 Tracking results for *(left)* pure 3D translation and *(right)* pure 3D rotation trajectories expressed in the world frame, with 'ZYX' Euler angles for rotation

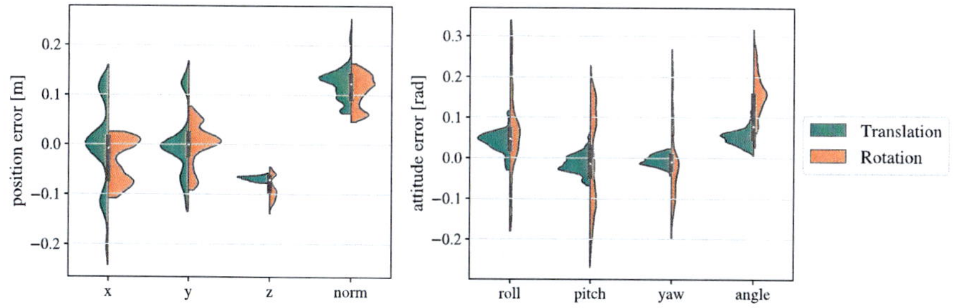

Fig. 6.4 Position *(left)* and attitude *(right)* error statistics are compared for pure translation and rotation trajectories, using tracking data shown in Fig. 6.3

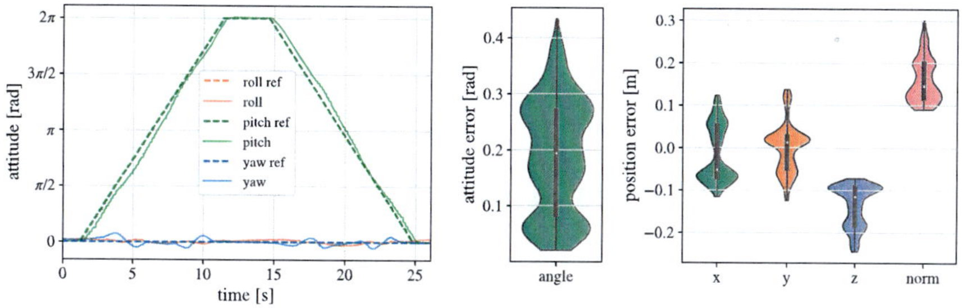

Fig. 6.5 Omnidirectional flight capabilities are shown with attitude tracking, and position and angular error distributions for a 360° rotation trajectory about the pitch axis. Values are in $\mathcal{F}_W$, with a translation offset to the nominal starting position

rotor airflow interference, and the limited representation of a linear propeller force model are likely the major effects in this case.

Errors related to tracking results in Fig. 6.3 are shown in Fig. 6.4, and are represented as violin plots showing the error distribution and statistics. Positional error for x and y linear tracking in the world frame are nominally within one centimeter of the target point. Deviations up to 24 cm for pure translational tracking are due to motion tracking delay of approximately 0.35 s, and the pure rotation tracking shows similar effects for angular error. Due to the constant height offset, the mean z position error is about 7 cm, resulting in a mean Euclidean position error of 12.2 cm. If the recorded position values are adjusted such that their mean value is the reference height, the resulting mean position error norm drops to 8.8 cm. Rotation error mean values for individual roll, pitch and yaw axes deviate from zero, due to combined model error effects including the center of mass position and tilt-arm zeo-position calibration. The resulting mean angular deviation is 4.5°.

Tracking results in Fig. 6.5 show the successful completion of two subsequent complete rotations about the reference pitch axis. Larger yaw deviations at 90° pitch and subsequent increments of 180° show the reduced actuation authority of the rank-reduced instantaneous allocation matrix and the associated tilt-motor delays when generating moment commands with a pure wrench model. The overall angular error

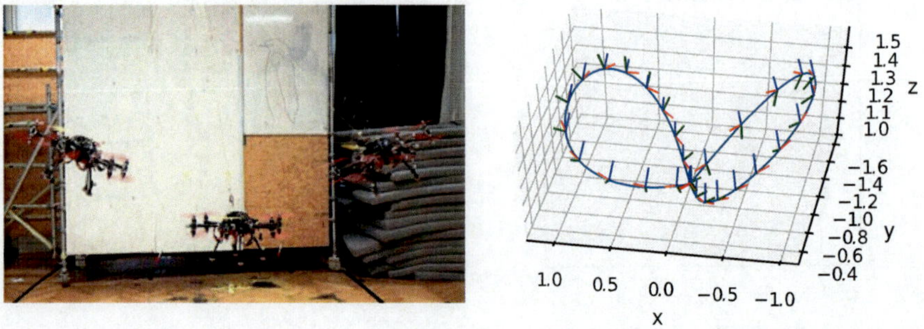

Fig. 6.6 Figure eight trajectory used for performance evaluation, (left) experimental flight and (right) reference trajectory plotted in $\mathcal{F}_W$. Frames indicate the base orientation at intermittent points spaced along the trajectory

mean is 11.1° over the entire dataset shown. As in the previous trajectories, the $\boldsymbol{e}_z^W$ positional error drives the overall position error, and can be corrected in future work with the use of an integral gain. Translational error in $\boldsymbol{e}_x^W$ and $\boldsymbol{e}_y^W$ directions remain below 12 cm.

B: Combined 6-DOF Trajectories

For the evaluation of arbitrary omnidirectional motion (simultaneous translation and rotation), we define a trajectory that covers a large part of the six DOF pose space. This trajectory consists of a figure-eight in position and yaw that rises in height and inclination at its apexes, and is parametric in its duration and maximum roll and pitch.

Figure 6.6 illustrates the figure-eight trajectory in a real flight experiment and plotted in $\mathcal{F}_W$ with some reference positions and attitudes of the base frame shown by coordinate frame axes. We use three different parametrizations of the figure-eight trajectory to evaluate controller performance, specifically:

4. Standard figure-eight with a maximum inclination of 30°, and duration of 29.4 s.
5. Fast figure-eight with a maximum inclination of 30°, and duration of **10.7 s**.
6. High-angle eight with a maximum inclination of **80**°, and duration of 29.4 s.

We first evaluate the performance of the standard figure-eight trajectory, which in its relatively slow movement examines the system's ability to simultaneously track changes in position and attitude. Compared to the previous experiments, these experiments use differential actuator allocation with the parameters presented in Table 6.5.

Overall tracking error statistics for each parametrized trajectory are presented in Table 6.6. In the following text, we present relevant tracking plots and error distributions for more detailed error analysis.

Table 6.6 Error statistics for standard, fast, and steeply inclined 6 DOF figure-eight tracking experiments. The mean of Euclidean error and total angular error are highlighted for each experiment

	Axis	x	y	z	Norm	Roll	Pitch	Yaw	Angle
Standard	Mean	−0.001	0.000	−0.033	**0.041**	−0.005	−0.020	−0.007	**0.066**
	std.	0.019	0.016	0.009	0.012	0.041	0.035	0.043	0.024
	RMS	0.019	0.016	0.034	0.042	0.042	0.041	0.044	0.070
Fast	Mean	0.001	−0.016	−0.034	**0.050**	−0.029	−0.100	−0.032	**0.128**
	std.	0.026	0.023	0.016	0.019	0.045	0.038	0.056	0.035
	RMS	0.026	0.028	0.038	0.054	0.053	0.107	0.064	0.133
Steep	Mean	−0.005	−0.006	−0.022	**0.046**	0.017	−0.028	−0.033	**0.101**
	std.	0.027	0.028	0.015	0.014	0.113	0.067	0.128	0.053
	RMS	0.028	0.029	0.027	0.048	0.114	0.073	0.132	0.114

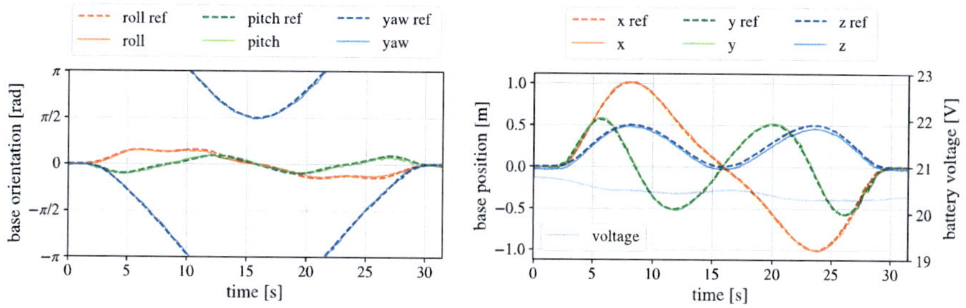

Fig. 6.7 Tracking results for the standard figure-eight trajectory, expressed in the world frame, with a translational offset to the starting position. The overall system voltage is plotted as an overlay on the position tracking plot

Tracking results of the standard trajectory are presented in Fig. 6.7. The plots show attitude tracking, wrapped in the range $(-\pi, \pi)$, and position tracking with the currently measured system voltage overlaid. Pose tracking plots indicate that the system can closely follow the assigned trajectory, with slight deviations visible particularly in the z position between 15 and 25 s.

The error distribution by axis, represented in Table 6.6, confirms our observation that $\boldsymbol{e}_z^W$ is the major axis for position error, suggesting model error in rotor thrust and inertial effects. An integrator is in effect for both position and attitude PID terms, but its effect varies with changing attitude, and as a result, does not completely negate the effects of model error. We notice a drop in voltage as the z position error grows, implying error in the voltage compensation model for rotor thrust mapping. Nevertheless, tracking results improve upon experiments in Sect. 6.2.1A for combined 6 DOF motion, with mean position error of 4.1 cm and mean angular error of 0.066 rad, corresponding to 3.8°.

To better understand the error distribution described in Table 6.6 over the three different trajectories, violin plots in Fig. 6.8 show the error distribution over each

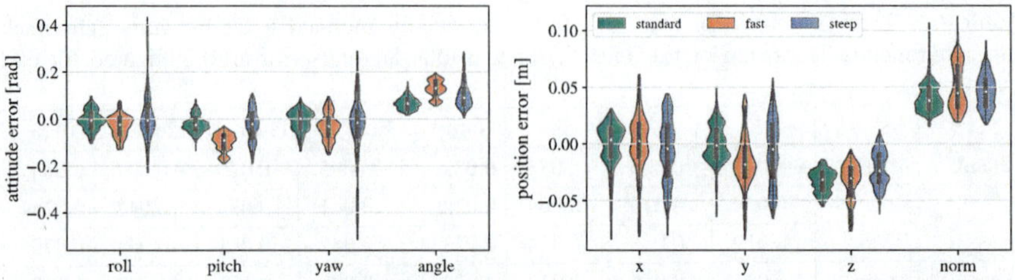

Fig. 6.8 Error distributions corresponding with error data in Table 6.6 are presented for standard, fast and steep variations of the figure-eight trajectory

axis, as well as total angular and Euclidean position errors. We notice a few general trends, such as the wider distribution of angular error for the steep trajectory variation. Since the inclination is significantly higher in this trajectory, larger airflow interaction is present which can lead to larger angular perturbations of the system. For the fast trajectory variation, the pitch axis distribution is consistently below the nominal value, and given that the $\boldsymbol{e}_x^B$ axis faces forward with motion along the figure-eight, this corresponds to a nose-up tendency. Model error is likely the cause of this phenomenon, since the feed-forward center of mass compensation term is highly sensitive to variation in the COM position. With faster movement, the system also encounters larger aerodynamic drag on the body, which is another contributing error factor. The nominal trajectory would minimize the body drag, as it orients the smallest projected area of the system in the direction of motion. However, as the pitch deviates due to inertial model error, a larger surface is exposed, creating positive feedback on the body drag, and requiring higher correction terms.

We look more closely at the error distribution for the steep trajectory condition, since this is the case that pushes our modelling assumptions the furthest. High roll and pitch angles lead to large interacting airflow streams which our current model neglects, and some actuators are pushed closer to the edge of the linear range for which our propeller model is valid. Figure 6.9 compares three trials of the steep figure-eight trajectory: with linear and angular integrators, and without integrators, including the case where the battery voltage drops very low. The addition of an integrator provides the best overall position and attitude tracking performance, though it results in the highest peak error in roll and yaw, which can be explained by integrator wind-up during a changing trajectory. In the absence of an integral term, several axes show increased deviation from zero-mean tracking, particularly the pitch, yaw, y and z axes. A bi-modal distribution on the x axis error represents an overshooting tendency, which is reduced with the addition of an integrator.

A trial with a very low battery exposes a performance limit of the system, when we consider the top right plot in Fig. 6.9. With higher roll and pitch angles, battery voltage drops with increased power-draw from the rotors for all trajectories. In a low-battery trial represented by the lowest voltage line on the plot, a critical point is reached around 19.2 V where the voltage-compensating rotor thrust model is outside of its range of validity, and as a result, a significant drop in altitude occurs at 25 s. As

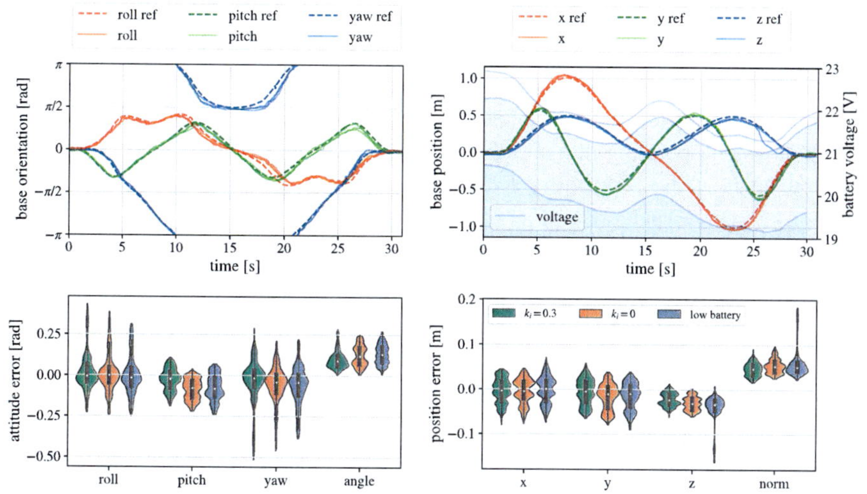

Fig. 6.9 Position and attitude tracking for three trials of a steep figure-eight trajectory, expressed in the world frame with a translation offset to the trajectory start point. Lower plots show the labelled error distributions for each trial. The low battery trial, which is also $k_i = 0$, corresponds to the position tracking plot that drops in elevation after 25 s

the pitch and roll recede, the voltage rises slightly again and the altitude recovers. This voltage-related altitude drop represents a detectable configuration-based failure mode. In future work, it would be interesting to optimize trajectories or modify the commanded orientation, considering feasibility based on the battery status and estimated voltage drop for a planned maneuver.

6.2.2 *Singularity Robustness and Secondary Tasks*

In certain conditions, our choice and realization of omnidirectional flying system negatively affect flight performance. The tilt-rotor morphology is subject to singularities as described in Sect. 4.1.3, and its construction is subject to tilt-angle limitations due to cable wind-up as described in Sect. 5.1.2. We can avoid and handle these cases through actuator allocation, taking advantage of the system's over-actuation. This series of experiments evaluates the ability for our two actuator allocation methods to handle singularity cases and manage cable wind-up. We judge performance based on trajectory tracking error, secondary task completion, maintained system stability, and force efficiency.

Experiments in group A evaluate singularity handling with an instantaneous allocation approach, where singularities are explicitly addressed and managed heuristically. Experiments in group B evaluate the differential allocation approach, which implicitly handles kinematic singularities, and can accommodate an unwinding task formulated in the actuation null space. As in the previous section, these allocation approaches are based on formulations in Sects. 4.1.3, 4.1.4 and parameters in Table 6.5.

We choose two trajectories, which compare performance of the actuator allocation methods when operating in a sustained singularity configuration, and transitioning through singularities. Trajectory details are as follows:

7. Cartwheel: Rotate to 90° roll and perform two rotations about $\boldsymbol{e}_z^B$ while performing a circle trajectory on the $\boldsymbol{e}_x^W$, $\boldsymbol{e}_y^W$ plane over a duration of 35.5 s.
8. Singularity translation: Rotate to 90° roll (two tilt-arms aligned with gravity in a kinematic singularity), then translate and change direction over a duration of 36.1 s.

The first trajectory performs a cartwheel motion, demonstrating repeated transition through the kinematic singularity, and the second commands a lateral translation (and rotation in series B) while in both a kinematic and rank-reduced singular state.

A: Explicit Singularity Handling and Unwinding

Tracking results are shown in Fig. 6.10. This series is using Proto^{V1} and instantaneous actuator allocation, with no integral term for the PID controller, resulting in a z position offset as seen previously in Sect. 6.2.1A, increasing for both trajectories as high roll and pitch angles are commanded. The commanded state of 90° roll corresponds to a state where control authority is at a minimum along $\boldsymbol{e}_z^B$ and about $\boldsymbol{e}_x^B$

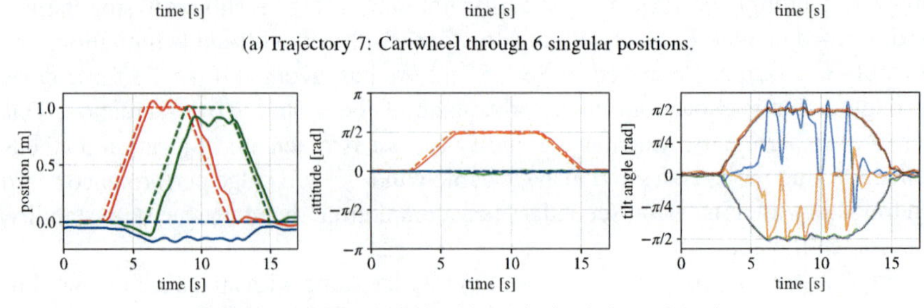

(a) Trajectory 7: Cartwheel through 6 singular positions.

(b) Trajectory 8: Translation in a kinematic singularity configuration.

Fig. 6.10 Experimental position tracking, attitude tracking, and commanded tilt angles for trajectories that enter singularity states, expressed in $\mathcal{F}_W$

and $\boldsymbol{e}_y^B$ axes, and instantaneous actuator allocation results in high tilt-arm velocities for small changes of the net aerodynamic actuation wrench in these directions.

We see the result of kinematic singularity handling (described in Sect. 4.1.3) when observing the commanded tilt-angles in Fig. 6.10. For the cartwheel trajectory, the kinematic singularity handling cone is activated with each vertical alignment of a tilt-arm, wherein the rotor forces of the corresponding propellers and tilt angles return to their zero (unwound) positions. As expected, position error grows in this configuration, since the ability to use upper and lower propellers for translation is sacrificed for singularity management. This condition is shown at its worst in the singular translation trajectory, maintaining the kinematic singularity condition while performing motion along $\boldsymbol{e}_x^W$ and $\boldsymbol{e}_y^W$. The plot of tilt angles shows that the singularity handling and unwinding command are effective in preventing high tilt angle velocities for low rotor force contributions and in preventing cable wind-up. As a result, the position error grows as the system is slow to converge to the reference trajectory, due to inhibited control authority.

In both cases which pass through known singularity regions, experimental results confirm stable tracking of the desired trajectory, despite reduced tracking performance. Unwinding of the cables is achieved only for arms that pass near a kinematic singularity, and wind-up is not prevented for pitch or roll flip trajectories as in Fig. 6.5. The approach in this case is heuristic and does not take full advantage of the system's over-actuation, motivating the use of a differential allocation method where the null space can be exploited directly.

B: Implicit Singularity Handling and Unwinding as a Secondary Task

A second series of experiments evaluates the singularity and unwinding performance of differential actuator allocation compared to direct actuator allocation experiments in Sect. 6.2.2A, and the new capabilities that come with differential allocation. The differential actuator allocation method inherently handles *kinematic singularities*, and facilitates unwinding of the tilt-arm cables as a complementary task in the actuation null space.

We noted in the previous section that the heuristic unwinding method only permits unwinding in a small region around a kinematic singularity. In the case of a flip about the pitch axis as shown in Fig. 6.5, a complete cable wind-up occurs about each arm although the kinematic singularity is avoided. Further rotation from this configuration results in a physical stop and subsequent crash since the cable-limitation is not expressly considered in the system model. We therefore take advantage of the system's over-actuation where unwinding cannot be solved by passing through kinematic singularities.

Using differential actuator allocation, the cable unwinding task is evaluated with three trajectories. The first is the figure-eight trajectory described in Sect. 6.2.1B, where the starting configuration includes four rotor groups wound by one rotation from the nominal position. The unwinding task is then activated at the start of the trajectory. The subsequent two trajectories perform complete rotations about the $\boldsymbol{e}_x^B$

Table 6.7 Error statistics for different trajectories while an unwinding task is assigned in the differential actuator allocation. The mean of Euclidean error and total angular error are highlighted for each experiment

	Axis	x	y	z	Norm	Roll	Pitch	Yaw	Angle
Singular	Mean	0.005	−0.027	−0.024	**0.049**	0.040	−0.083	−0.019	**0.114**
	std.	0.026	0.020	0.014	0.013	0.037	0.063	0.036	0.047
	RMS	0.027	0.034	0.027	0.051	0.054	0.104	0.041	0.123
Cartwheel	Mean	0.004	−0.005	−0.024	**0.050**	−0.040	−0.030	0.013	**0.153**
	std.	0.030	0.029	0.017	0.012	0.064	0.133	0.102	0.056
	RMS	0.031	0.030	0.029	0.052	0.075	0.136	0.103	0.163
Standard	Mean	0.001	−0.006	−0.037	**0.044**	0.020	−0.022	−0.047	**0.105**
	std.	0.017	0.016	0.012	0.011	0.062	0.068	0.047	0.042
	RMS	0.017	0.017	0.039	0.046	0.065	0.071	0.066	0.113
Pitch flip	Mean	0.007	0.007	−0.021	**0.045**	0.016	0.032	0.112	**0.197**
	std.	0.035	0.012	0.023	0.021	0.031	0.124	0.121	0.078
	RMS	0.036	0.014	0.031	0.049	0.035	0.128	0.165	0.212
Roll flip	Mean	−0.013	−0.010	−0.019	**0.041**	0.008	−0.027	−0.028	**0.072**
	std.	0.007	0.020	0.028	0.014	0.042	0.064	0.033	0.057
	RMS	0.015	0.022	0.034	0.043	0.042	0.069	0.043	0.092

and $\boldsymbol{e}_y^B$ axes starting with nominally unwound rotor groups, in a more realistic case where the trajectory itself creates the need for unwinding. Detailed descriptions of the additional trajectories are as follows:

9. Figure eight with a maximum inclination of 30°, with four rotor groups starting at $\alpha_i = 2\pi$, over a duration of 29.4 s.
10. Pitch flip: complete rotation about the $\boldsymbol{e}_y^B$ axis over a duration of 8 s.
11. Roll flip: a complete rotation about the $\boldsymbol{e}_x^B$ axis over a duration of 16 s.

Overall tracking error statistics for each of trajectories 7–11 are presented in Table 6.7. Corresponding error distributions are shown as violin plots in Fig. 6.11, including the error distribution over each axis, and the total angle and Euclidean position errors. Singular translation and cartwheel trajectories have the highest overall position error, which can be explained by extended time spent in singular configurations with relatively low actuation authority. The overall system performance is an improvement in position and attitude tracking performance upon the instantaneous allocation method.

When comparing the results of the cartwheel trajectory shown in Fig. 6.12 to those of using instantaneous allocation in Fig. 6.10, a major difference is visible in the tilt angle commands. The heuristic unwinding method causes tilt angles to snap to zero when the singularity handling is active and the vertical rotors lie within this deadzone. For differential allocation, preferring low velocity of tilt-angle speeds drives a smooth transition of tilt-angles through the singularity, resulting in much better tracking performance in these conditions.

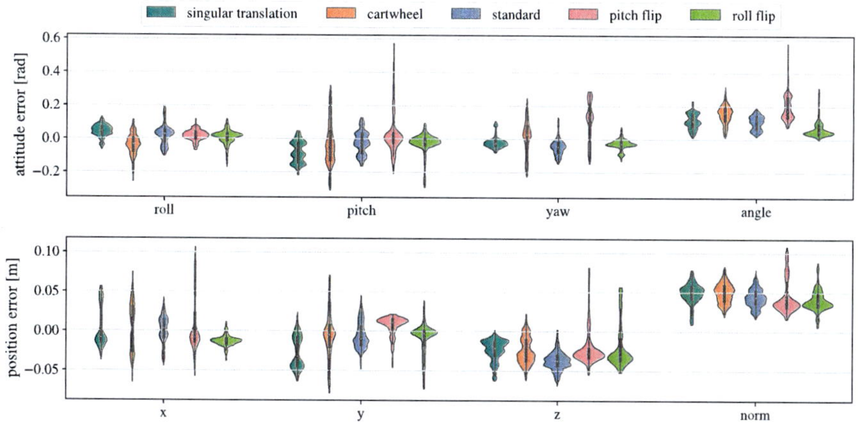

Fig. 6.11 Error statistics for different trajectories during an unwinding task corresponding to error data in Table 6.7. All axes are in the world frame, with attitude expressed as yaw-pitch-roll Euler angles

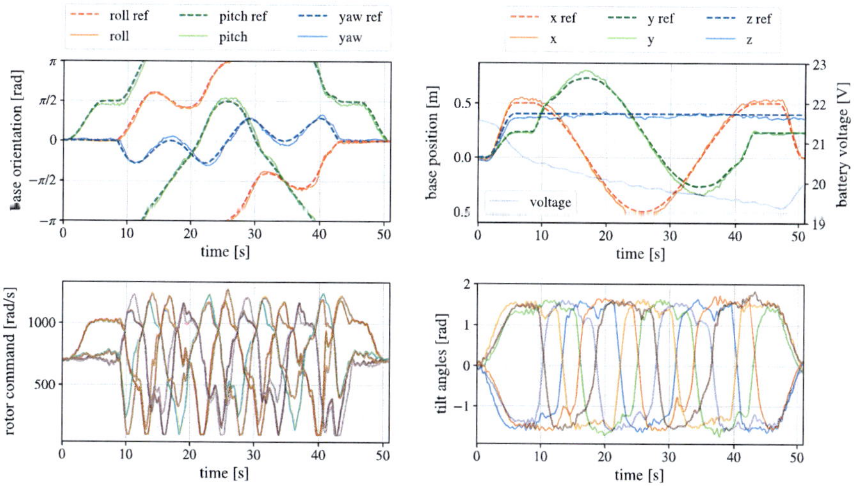

Fig. 6.12 Trajectory 7: Cartwheel with differential allocation. Top: attitude and position tracking plots with overlaid system voltage. Bottom: rotor speed and tilt angle commands for all actuators

Similar to the cartwheel trajectory, the singular translation trajectory shown in Fig. 6.13 does not force tilt-angles to zero to prevent cable wind-up, which was with instantaneous allocation in Fig. 6.10. Instead, the inherent cable wind-up prevention allows optimal use of the vertically oriented rotor groups according to their current tilt-orientation, increasing the control authority and reducing position tracking error compared to instantaneous actuator allocation. While the rotor commands shown in the lower left plot of Fig. 6.13 show that upper and lower rotors drop significantly in contribution, they are still in active use.

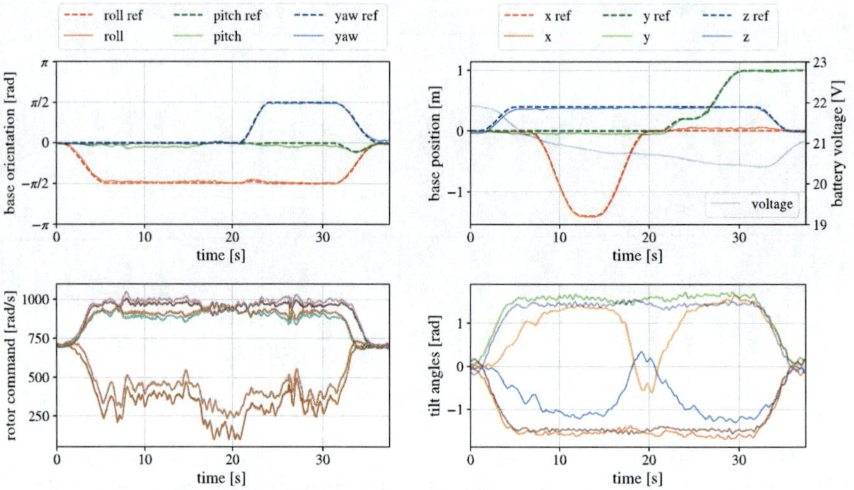

Fig. 6.13 Trajectory 8: Singular translation with differential allocation. Top: attitude and position tracking plots with overlaid system voltage. Bottom: rotor speed and tilt angle commands for all actuators

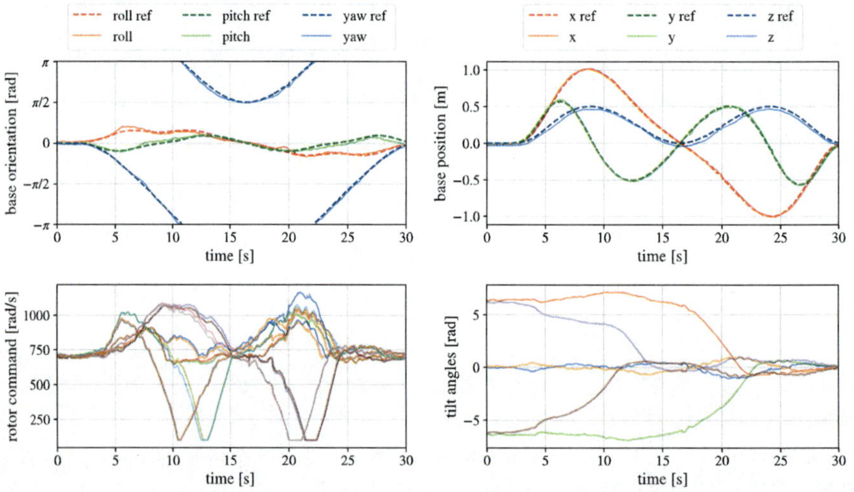

Fig. 6.14 Trajectory 9: Standard 6 DOF trajectory with 4 wound rotors and differential allocation. Top: attitude and position tracking plots. Bottom: rotor speed and tilt angle commands for all actuators

Figure 6.14 presents tracking results for a standard 6 DOF figure-eight trajectory where four actuators start in a wound state. The results demonstrate that the system can achieve good tracking while unwinding dangerously wrapped cables about the tilt-arms without entering the kinematic singularity. Unwinding can be seen clearly in the lower plots, where rotor commands are at their minimum when completely

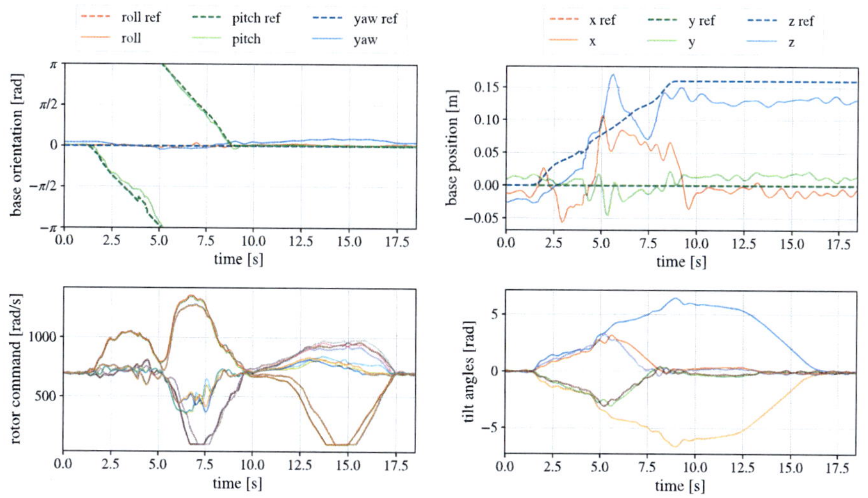

Fig. 6.15 Trajectory 10: 360° pitch flip with differential allocation. Top: attitude and position tracking plots. Bottom: rotor speed and tilt angle commands for all actuators

inverted from the contributing direction during an unwinding event. The unwinding timing is never prescribed, instead occurring as a result of the trajectory, the task specification and allocation tuning parameters. In this flight case, unwinding occurs sequentially at increased trajectory roll and pitch. In the first loop of the figure-eight, two arms are unwound completely. The opposite two arms are unwound in the second loop.

Figure 6.15 presents the first case where the need for unwinding is driven by the motion of the base itself. In Fig. 6.5, this trajectory causes wind-up of all tilt-arms since heuristic unwinding conditions (entering the kinematic singularity region) are not achieved. A slight rise in height is commanded for this and the following two trajectories, due to the current planner's requirement for a small translation to generate a trajectory. This vertical motion is small and slow, and its effect is insignificant compared to the rotational motion.

The four tilt-arms which are not collinear with the $\boldsymbol{e}_y^B$ unwind during the second half of the rotation. The remaining two rotor groups unwind several seconds after the completion of the flip during flat hover. Again, the choice of allocation parameters in Table 6.5 influence the unwinding behaviour, and they were carefully chosen to balance timely unwinding with system stability and tracking performance.

Tracking results for a 360° flip about the roll axis are shown in Fig. 6.16. Compared to the pitch flip, two tilt arms in this trajectory pass through the kinematic singularity twice, performing an unwinding operation on the second pass. The remaining tilt-rotors do not unwind during the trajectory, but instead unwind in pairs once the system has returned to static hover. This behaviour is once again explained by the choice of allocation parameters, and compared to the mid-flip unwinding performed during the pitch flip, the balance of rotor effort and unwinding gain did not result

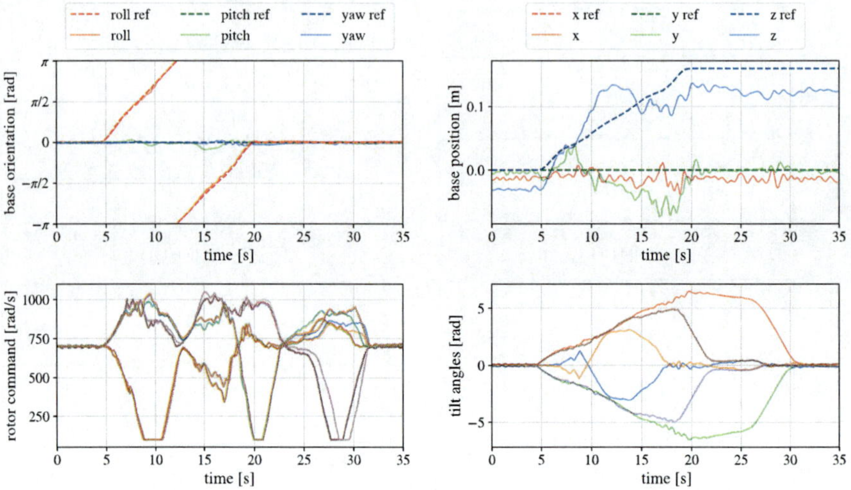

Fig. 6.16 Trajectory 11: 360° roll flip with differential allocation. Top: attitude and position tracking plots. Bottom: rotor speed and tilt angle commands for all actuators

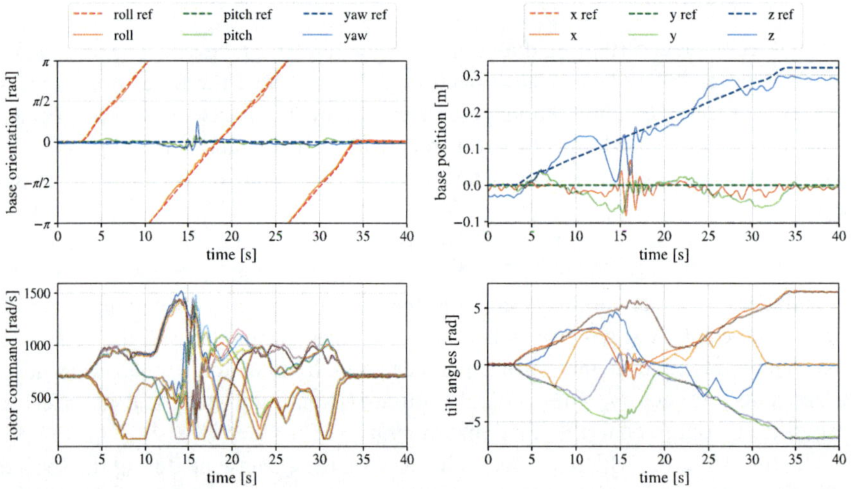

Fig. 6.17 720° roll flip with differential allocation. Top: attitude and position tracking plots. Bottom: rotor speed and tilt angle commands for all actuators

in unwinding until the trajectory was complete. We notice larger tracking errors in pitch at approximately the 9 and 16 s marks, corresponding to the rank reduction singularity where pitch authority is instantaneously reduced.

After the unwinding delay shown in the previous roll flip experiment, we test the method's ability to handle a 720° roll flip, which would lead to system failure for the instantaneous allocation's heuristic unwinding case. Tracking results in Fig. 6.17

show that a second rotation causes the actuator allocation advance unwinding events, preventing catastrophic cable wind-up. Additional attitude and position error around 16 s just following the second kinematic singularity show the corresponding reduction in tracking performance with an increase in required unwinding.

Pose tracking results for all differential allocation experiments show satisfactory performance, but rely on careful selection of gains. Since rotor speed commands have no upper bound in the optimization, a more aggressive choice of gains could result in motor saturation and failure to maintain hover. Improvements to the presented differential allocation method would be the incorporation of these constraints, and guarantees on the system's ability to maintain hover. Knowledge of the future trajectory in a receding horizon framework would also provide an interesting advantage for the timing of unwinding events.

6.2.3 *Efficiency Evaluation*

Apart from trajectory tracking, flight efficiency is another key performance factor for omnidirectional free flight. To evaluate the system's efficiency, we recall the force efficiency metric, η_f, presented in Eq. 2.12, as the magnitude of the net actuation force divided by the sum of all rotor speeds, which lies in the range [0, 1].

We further consider the practical metrics of total power consumption and power efficiency, η_P, where the latter is defined relative to the best-case power consumption p_h of horizontal hover,

$$\eta_P = \frac{v_{\text{sys}} \Sigma_{i=1}^{12} i_i}{p_h} \in [0, 1] \tag{6.2}$$

where i_i is the measured current for the ith rotor group and v_{sys} is the total system voltage.

Theoretical total power, power efficiency and wasted force for each hover orientation are shown in the upper plots of Fig. 6.18, projected onto the unit sphere and coloured by magnitude. Lower plots show measured values from experimental results with instantaneous allocation, using measured PWM-to-current correlation and flight-test voltage logs to find the power consumed. Experimental results agree well with the ranges predicted on the theoretical envelopes, with the theoretical value acting as an upper bound. These results confirm that highly efficient flight configurations are achieved in specific orientations, while the system remains pose-omnidirectional with varying degrees of efficiency.

In the case of differential allocation, efficiency is traded off when there is a need for cable unwinding away from a kinematic singularity. Figure 6.19 compares the experimentally achieved force efficiency index η_f with the ideal maximum achievable efficiency, which is computed assuming stationary hover in the reference attitude and perfect trajectory tracking.

For the singular translation and cartwheel trajectories, efficiency values are close to theoretically optimal values, with a general loss in efficiency due to control error

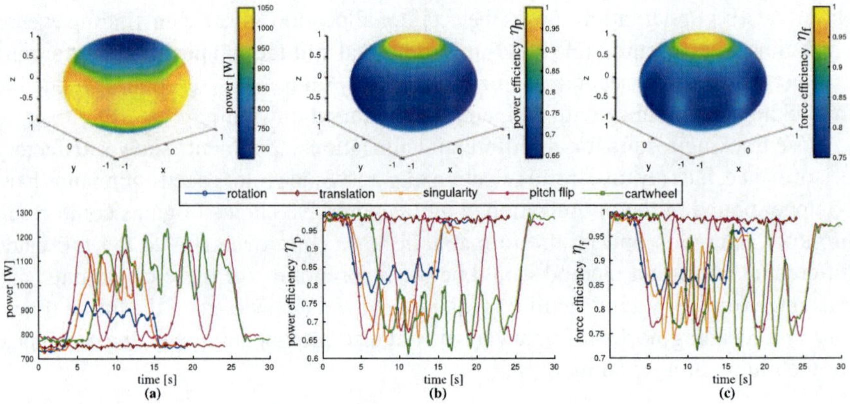

Fig. 6.18 **a** Power consumption, **b** power efficiency and **c** wasted force index for theoretical hover, and measured values from tested trajectories. Upper plots show theoretical values projected onto the unit sphere of omnidirectional hover orientations, coloured by magnitude

corrections. In the case of the singular translation, any activation of vertically aligned rotors to improve control correction detracts from the flight efficiency, since these forces have no contribution against gravity.

In the case of the cartwheel, the theoretical η_f does not form a clean upper bound for the experimental data. In general the efficiency is offset in time from the theoretically perfect one, indicating an advanced trajectory performance. We expect that this is due to model error relative to the feed-forward trajectory acceleration terms, and relatively low correction control authority of the current tilt-arm configuration. In some cases where the target trajectory is the least efficient configuration, trajectory deviations can even improve efficiency.

For trajectories 9–11, the standard figure-eight, pitch flip and roll flip, we notice large deviations from the theoretical maximum efficiency. These periods in time correspond with unwinding events that can be seen in Figs. 6.14, 6.15, 6.16. In the case of the standard figure-eight trajectory, efficiency is reduced by 38% from the nominal value at 7.5 s and drops as low as 53% during the pitch flip.

Overall, the differential actuator allocation demonstrates a great advantage over the instantaneous allocation approach in its ability to prevent damage to the system due to cable wind-up on any flight trajectory. This unwinding capability comes with the drawback of reduced flight efficiency, sometimes with inopportune timing, and no guarantee of flight feasibility. Both of these elements are good candidates for future extensions of differential allocation to consider a receding horizon of the trajectory, and guarantee stable flight. The remaining advantage of the instantaneous allocation method is its computational compactness, requiring much less compute to arrive at a good solution for flights that do not encounter cable wind-up. This lightweight method is deployable on low-power microcontrollers, and is more accessible to the general public in terms of implementation and tuning.

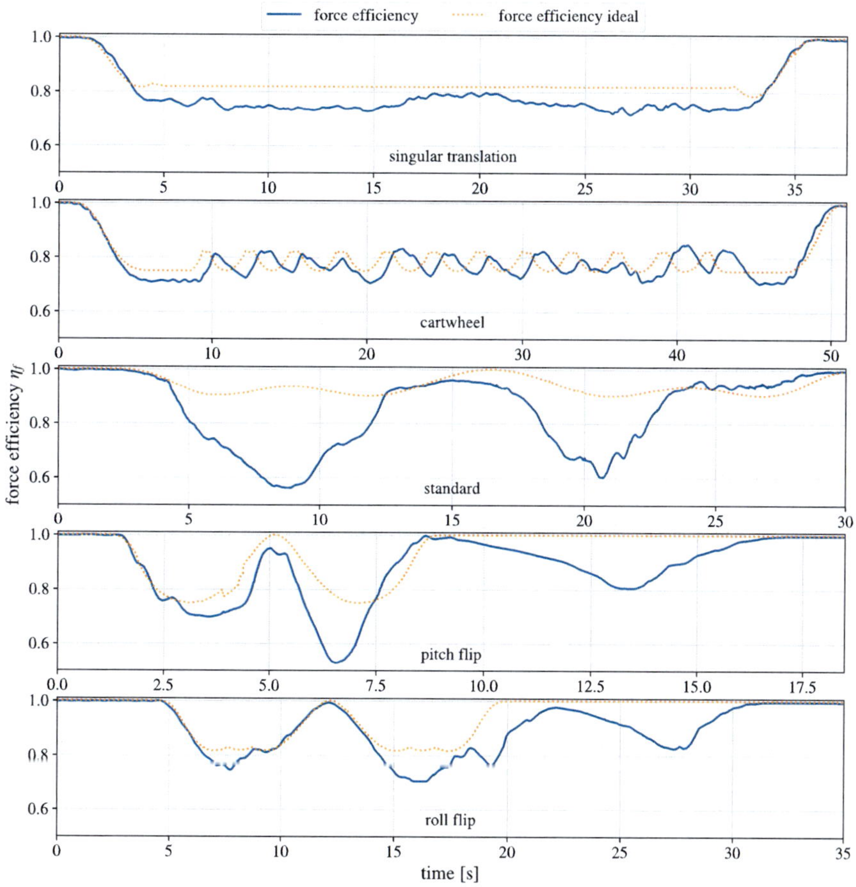

Fig. 6.19 Force efficiency for differential allocation flights compared to theoretical efficiency for trajectories 7–11. The dotted orange line corresponds with the force required to maintain hover in the prescribed orientation if perfect tracking is achieved, with the most efficient tilt-angle configuration. The blue line represents the force efficiency computed according to the metric $\eta_f = \frac{\|\sum_i f_{R_i}\|}{\sum_i |f_{r_i}|}$

6.3 Omnidirectional Aerial Physical Interaction

Based on the system validation of an omnidirectional tilt-rotor in Sect. 6.2, this section contains a suite of experiments that demonstrate aerial physical interaction using the same system with a rigidly mounted end-effector. Experiments are conducted using Proto^{V1} and Proto^{V2}, where we evaluate the performance of axis-selective impedance control and its extension to direct interaction force control (described in Sects. 4.2.2, 4.2.3) using instantaneous actuator allocation (described in Sect. 4.1.3). Experiments evaluate contact wrench estimation and tracking, disturbance rejection, and combined

Table 6.8 Experimental setup for aerial physical interaction tests, for axis-selective impedance control (ASIC) and hybrid force-impedance control, a hybrid of direct interaction force and variable axis-selective impedance control (VASIC)

Experiment	Prototype	Controller	Test env.	State est
Section 6.3.1 Section 6.3.2A Section 6.3.4A Section 6.3.6A	Proto^{V1}	ASIC	large arena	Vicon-MSF
Section 6.3.5A	Proto^{V1}	ASIC	archway	VIO-MSF
Section 6.3.2B Section 6.3.4B	Proto^{V2}	hybrid force/VASIC	small arena	Vicon-MSF
Section 6.3.3 Section 6.3.5B	Proto^{V2}	hybrid force/VASIC	large arena	Vicon-MSF

force and pose interaction tasks such as aerial drawing and NDT contact inspection. We perform the following series of experiments:

- Section 6.3.1 presents estimates and measurements of the **external contact wrench** at the end-effector point, including the tuning of a momentum-based wrench estimator, the filtering of force sensor readings, and the comparison of these methods.
- Section 6.3.2 evaluates the system response to an applied disturbance with fixed and variable axis-selective impedance control, demonstrating **disturbance rejection** behaviour, and the effect of selected virtual inertia values.
- Section 6.3.3 demonstrates **direct force control** during interaction tasks, and performance robustness to planning errors.
- Section 6.3.4 presents **push-and-slide** experiments to repeatable pose tracking when interacting with a planar surface, while rejecting disturbances due to surface friction. Both impedance control and direct force control are evaluated and compared.
- Section 6.3.5 demonstrates interaction with **inclined and non-planar surfaces**, for both impedance and direct force control, including the case of fully on-board state estimation with local planning.
- Section 6.3.6 validates the system's viability as an autonomous tool for **NDT contact inspection** of infrastructure.

The experimental setup for aerial physical interaction test flights is described in Table 6.8, grouped by prototype system, controller, test environment and state estimator. Most experiments take place using Vicon-MSF state estimation, where the system is also connected by a light safety tether. Section 6.3.5A is the only experiment where VIO-MSFstate estimation is used, and this experiment along with Sect. 6.3.5B does not have an attached safety tether.

Controller Parameters

Controller parameters and gains used for physical interaction experiments are presented in Table 6.9, any variations from these parameters are stated in the corresponding experiment subsection. Parameters used for the PID component of the impedance

Table 6.9 Controller parameters used for aerial physical interaction experiments, where impedance control parameters correspond with methods from Sect. 4.2.2 and hybrid control parameters correspond to methods from Sect. 4.2.3

Parameter	Value
$m_{v,\text{free}}$	5.0
$m_{v,\text{wall}}$	0.25
$\mathcal{I}_v$	5.0

(a) Impedance control.

Parameter	Value
k_{p_f}	0.1
k_{i_f}	1.0
k_{d_f}	0.0

(b) Direct force control.

Parameter	Value	Unit
$e_{\min}$	0.15	[m]
$e_{\max}$	0.35	[m]
$d_{\min}$	0.02	[m]
$d_{\max}$	0.2	[m]
c_λ	0.01	-

(c) Hybrid control.

control are taken from Table 6.5. For momentum-based wrench estimation, diagonal values of $\boldsymbol{K}_I$ are set to unity unless otherwise stated.

6.3.1 Contact Wrench

We evaluate measured and estimated contact wrenches, which are used for interaction control experiments in the remainder of this section. Measurements from 6-axis force-torque sensors are used for ground truth and direct force control, and here we evaluate their raw and filtered signals, as well as bias. We then evaluate the performance of the momentum-based wrench estimator used in impedance control, and consider the effects of tuning the integral term.

Force Sensor Noise and Error

Aerial physical interaction experiments with direct force feedback use measurements from a 6-axis force-torque sensor mounted close to the end-effector point. Force sensors such as these are a well-established tool for accurate and precise force feedback, but are subject to new challenges when mounted on rotary-wing systems. High-frequency propeller motion and the resulting periodic downwash on supporting rotor group arms generates vibrations that affect the entire system. These vibrations are amplified when problems arise in the rotor group mechanics, such as damaged propellers and rotor bearings, or badly balanced propellers. Strain gauge force sensors such as the one we use are sensitive to high-frequency vibrations, which generate noise in the readings. We can address these vibrations with a low-pass filter.

We design an experiment to evaluate the additional noise of the force sensor mounted to a rotary-wing system. With one sensor mounted below the end-effector,

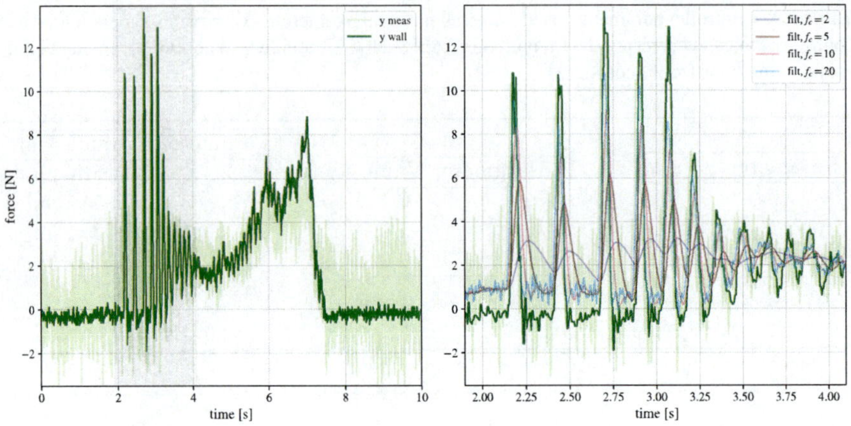

Fig. 6.20 Measurements from an on-board force-torque sensor at the end-effector are compared to values from a wall-mounted force sensor of similar properties, where the system is in steady contact with the wall between 3.5 and 7 s. The grey area on the left plot (bouncing during initial contact) is magnified on the right, where we see the effect of different cut-off frequencies for a low-pass Butterworth filter applied to the on-board force signal

and another mounted to a wall aligned the $\boldsymbol{e}_y^W$ plane, we plan a trajectory that brings the end-effector in contact with the wall sensor. Figure 6.20 compares the data collected from both sensors, where we see the noise of the on-board sensor is much more significant. When not in contact and the aerial system is in static hover, the onboard force sensor data has a standard deviation of 1.59 N compared to 0.18 N for the wall-mounted sensor.

To filter the force sensor signal, we compare the result of applying a 2nd-order low-pass Butterworth filter with cut-off frequencies of 2, 5, 10, and 20 Hz. Our goal is to capture transients in force readings with reasonably low delay, while eliminating high measurement peaks and removing high-frequency noise. In the right plot of Fig. 6.20 we see the effect of the different filters during the initial contact phase, where the end-effector bounces against the wall until steady contact is maintained at around 3.5 s. The 2 Hz cut-off frequency does not adequately capture the signal, introducing large delays that could lead to instability in the force feedback loop. A cut-off frequency of 10 Hz still shows noise in the troughs of the contact bumps, with the 20 Hz filter exaggerating the noise further. While these fluctuations are comparable to the wall-mounted sensor's raw reading, we prefer to have a smoother trajectory for force feedback in interaction. The 5 Hz filter provides an acceptable compromise, presenting a smooth signal with 0.02 s delay for reaching the high peak and 0.1 s delay for the signal to reach its low point during the contact transition. This filter yields smooth force measurements, at the sacrifice of a small delay in resolving transients. This latency is reasonable in the context of surface inspection tasks with our proposed controllers, which will be demonstrated in the remainder of this section.

Sensor noise could be reduced with additional design considerations. To achieve good model fidelity, the system design is very stiff with rigidly mounted carbon fibre and aluminium components, which allows high-frequency vibrations to propagate. Future work that incorporates an appropriately damped mount for the force sensor would likely improve the signal and response of force feedback.

A second new challenge that rotary-wing vehicles present for force sensors is the variable cooling effect of the propeller down-wash. Since these sensors are typically stationary, zeroing the reading once they have turned on and gained a steady state internal temperature is sufficient for a reliable reading over time. The aggressive cooling effects of propellers present a challenging temperature calibration problem, which changes depending on the configuration of our tilt-rotor system. The result is a changing force bias that evolves over the first few minutes after take-off, already a substantial amount of our total flight time. To address this issue during experiments, we zero the force sensor before each contact experiment after taking off and hovering for several minutes. Shielding or insulating the sensor casing from the surrounding airflow is another strategy, but may result in dangerous overheating in the sensor electronics. A robust solution will include temperature calibration which carefully models the effects of temperature transients on sensor readings. This approach remains a topic for future work.

Momentum Based Wrench Estimate

We use a momentum-based wrench estimator derived in in Eq. 4.46 as a fundamental element of impedance control. In this section, we evaluate the performance of this model-based estimator compared to ground truth force data in experimental flights.

Force ground truth data were collected from a 6-axis Rokubi 210 force-torque sensor with its surface aligned with the $\boldsymbol{e}_x^W$-plane, rigidly mounted to a wall. The force sensor recorded measurements at 800 Hz, with a resolution of 0.1 N. The system follows a trajectory along the $\boldsymbol{e}_x^W$-axis which goes in and out of contact with the force sensor. The system goes into contact twice, first with a low, then higher apparent mass, corresponding with low and high interaction forces.

Force data over the trajectory is shown in Fig. 6.21, where the wall-mounted force sensor data is smoothed with a 5th order low-pass Butterworth filter with a

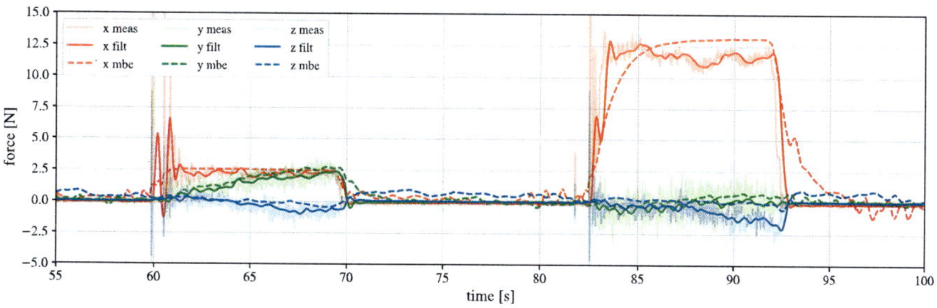

Fig. 6.21 Momentum-based external contact wrench estimates (mbe) are plotted against raw (meas) and filtered (filt) force sensor data, as the aerial system contacts a force sensor mounted to a wall

Table 6.10 The RMSE of force estimates

f_x (N)	f_y (N)	f_z (N)
1.24	0.38	0.63

cut-off frequency of 10 Hz. The filtered and raw ground truth data are plotted along with the momentum-based estimates, and presented values are offset by steady-state bias values measured during static hover. In general, force magnitudes match the ground truth measurements well. The RMSE of the force estimate compared to filtered ground truth are presented by axis in Table 6.10. In the direction of contact, an error of 1.24 N is due largely to a slow transient response to the contact transition, influenced by the choice of tuning parameter $\boldsymbol{K}_I$. Due to this slower response, the force bouncing during initial contact is not resolved, which is a preferred behaviour in our case for contact transition, preventing instability caused by oscillating forces and actuation delays. Torque error was not evaluated at this stage since we consider a point contact in experiments, but we expect it to behave similarly given the controller performance in the following sections. Experiments could conclusively measure this by for instance fixing a magnetic attachment to a flat surface at the end-effector, and generating torques about the rotationally constrained contact point.

Wrench Estimator Tuning

The momentum-based estimator is a model-based first-order low-pass filter with one tuning parameter: the integral gain. This value was tuned based on the behaviour of the system over different free-flight and contact interaction experiments. The choice of parameter was based on corresponding tracking performance and visual flight stability more than accurate representation of the contact force.

In Fig. 6.22 we compare the effect of the momentum-based wrench estimator for different values of k_I, where $\boldsymbol{K}_I = k_I \mathbb{1}_6$. These trials use impedance control which relies on the wrench estimator values during interaction. Plotted against a filtered

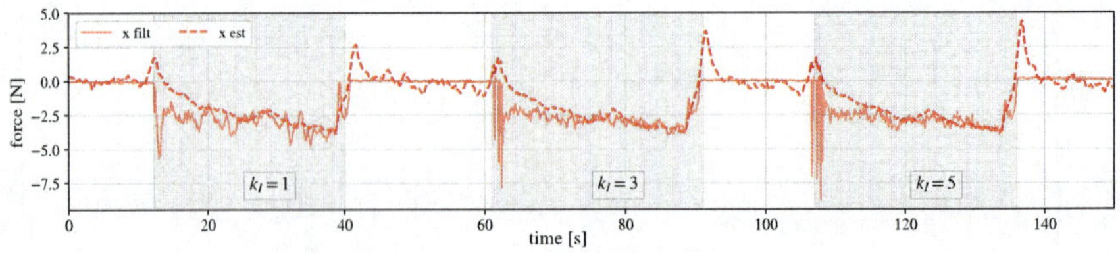

Fig. 6.22 The effect of the wrench estimator integrator is shown in three drawing trials using impedance control. Force is expressed in the world frame, where x is the normal direction of the whiteboard. Highlighted sections indicate regions of contact, and k_I values are multiplied with identity to obtain the gain matrix $\boldsymbol{K}_I = k_I \mathbb{1}_6$

on-board force sensor, we see that the wrench estimate fluctuates by around 1 N of the measured value on all axes due to model error, and deviates more dramatically on the contact axis during the transition into contact. We notice that a higher k_I value corresponds with slightly faster convergence to the contact reference. An increased k_I also reveals the more critical effect of significant bouncing during the contact transition, as shown by high force peaks on the $\boldsymbol{e}_x^W$ axis as the system gains contact. We can explain this phenomenon by considering the low bandwidth of the actuation wrench. A more aggressive gain directly contributes to the command wrench, but when its response frequency cannot be achieved by the aerodynamic control wrench, we see the dramatic effect of bouncing due to actuation delay. Choosing $k_I = 1$, we create a well-behaved interaction system as long as the system is relatively slow-moving. A higher performance solution will be to increase the actuation bandwidth of the system and raise the estimator gains, or incorporate the system dynamics as a component of an external disturbance observer.

6.3.2 *Disturbance Rejection*

Whether or not the system is in contact with the environment, external disturbances can significantly reduce task performance. Our axis-selective impedance control scheme combats these effects by estimating and rejecting disturbances due to wind gusts, contact and model error. The axis selected for interaction, on the other hand, is compliant to contact forces as well as to other disturbing forces.

In the first series of experiments, we evaluate the system response to an external disturbance for **axis-selective impedance control** when in free flight, evaluating the pose tracking for fixed axis-selective virtual inertia values. We follow this with a series of tests that evaluate **variable axis selective impedance control** using distance feedback in the end-effector direction, and evaluate the system behaviour when contact is expected. For these experiments, the end-effector is forward-facing, having its $\boldsymbol{e}_z^E$ axis aligned with the platform's $\boldsymbol{e}_x^B$ axis.

A: Axis-Selective Impedance Control

We evaluate the behaviour of the system with different selective apparent inertia values, demonstrating the ability to reject large disturbances in certain directions while exhibiting compliance in others. The system is commanded to hold a reference pose 1 m above the ground in free flight. A cord is tied to the end-effector tip of the rigid manipulator arm, which is aligned with the $\boldsymbol{e}_x^B$ ($\boldsymbol{e}_z^E$) axis. The other end of the cord is pulled manually to generate an external wrench. The tests are performed with Proto^{V1} in the small flight arena using Vicon external motion capture data fused with IMU for state estimation.

Table 6.11 Virtual mass parameters in $\mathcal{F}_B$ for axis-selective impedance control disturbance trials, where the virtual mass matrix is constructed as diag($[m_{v,x}\ m_{v,y}\ m_{v,z}\ \mathcal{I}_{v,x}\ \mathcal{I}_{v,y}\ \mathcal{I}_{v,z}]^\top$)

Test	$m_{v,x}$	$m_{v,y}$	$m_{v,z}$	$\mathcal{I}_{v,x}$	$\mathcal{I}_{v,y}$	$\mathcal{I}_{v,z}$
1	**0.25**	**0.25**	1.0	1.0	1.0	1.0
2	**0.1**	**0.1**	1.0	1.0	1.0	1.0
3	**5.0**	**5.0**	1.0	1.0	1.0	1.0
4	5.0	5.0	5.0	5.0	5.0	**0.25**
5	5.0	5.0	5.0	5.0	5.0	**5.0**

Virtual mass parameters for each trial are listed in Table 6.11. In tests 1 through 3, with results shown in Fig. 6.23, two pulls of the rope are made for each set of apparent inertia parameters, approximately along the negative $\boldsymbol{e}_x^B$ axis. In tests 4 and 5, with results shown in Fig. 6.24, a pull force is applied horizontally perpendicular to the fixed arm axis to generate a torque about the $\boldsymbol{e}_z^B$ axis.

Tests 1 and 2 show similar results: a compliant response to a disturbance force in the direction of pull. Apparent mass values in $\boldsymbol{e}_x^B$ and $\boldsymbol{e}_y^B$ axes are lower than the actual system mass, meaning that force disturbances in these directions will be tracked in the controller, while the PD component simultaneously tracks the reference trajectory. Results in test 2 show a larger movement in response to a smaller applied force in low impedance directions, relative to test 1. The remaining degrees of freedom have high apparent inertia values, actively rejecting detected disturbances to track the reference trajectory. In test 3, apparent mass along $\boldsymbol{e}_x^B$ and $\boldsymbol{e}_y^B$ are set to 5 times the system mass and inertia. Results show positional movement of less than 0.3 m under a lateral disturbance force of 25 N, demonstrating an ability to actively reject large force disturbances.

Tests 4 and 5 compare the response to a torque disturbance about the $\boldsymbol{e}_z^B$ axis with apparent rotational inertia less than, and greater than the system inertia. In test 4, three pulls are made, targeting rotation about the $\boldsymbol{e}_z^B$ axis, and rotational compliance is clearly shown in yaw in the attitude tracking plot. While some additional torques are generated around the remaining axes, these are actively rejected by the controller. In test 5, the system counteracts a rotational torque of 3N m magnitude, reducing the yaw deviation to 0.5 rad. High apparent mass in all directions successfully rejects forces up to 8 N with a translational deviation of less than 0.1 m.

B: Variable Axis-Selective Impedance Control

While axis-selective impedance control is a simple and practical solution for simultaneous interaction and disturbance rejection, it is always compliant to disturbance forces along the contact axis. With **variable** axis-selective impedance control, we allow the virtual mass along the contact axis to change based on a distance sensor measurement, so that omnidirectional disturbance rejection is enforced when we are

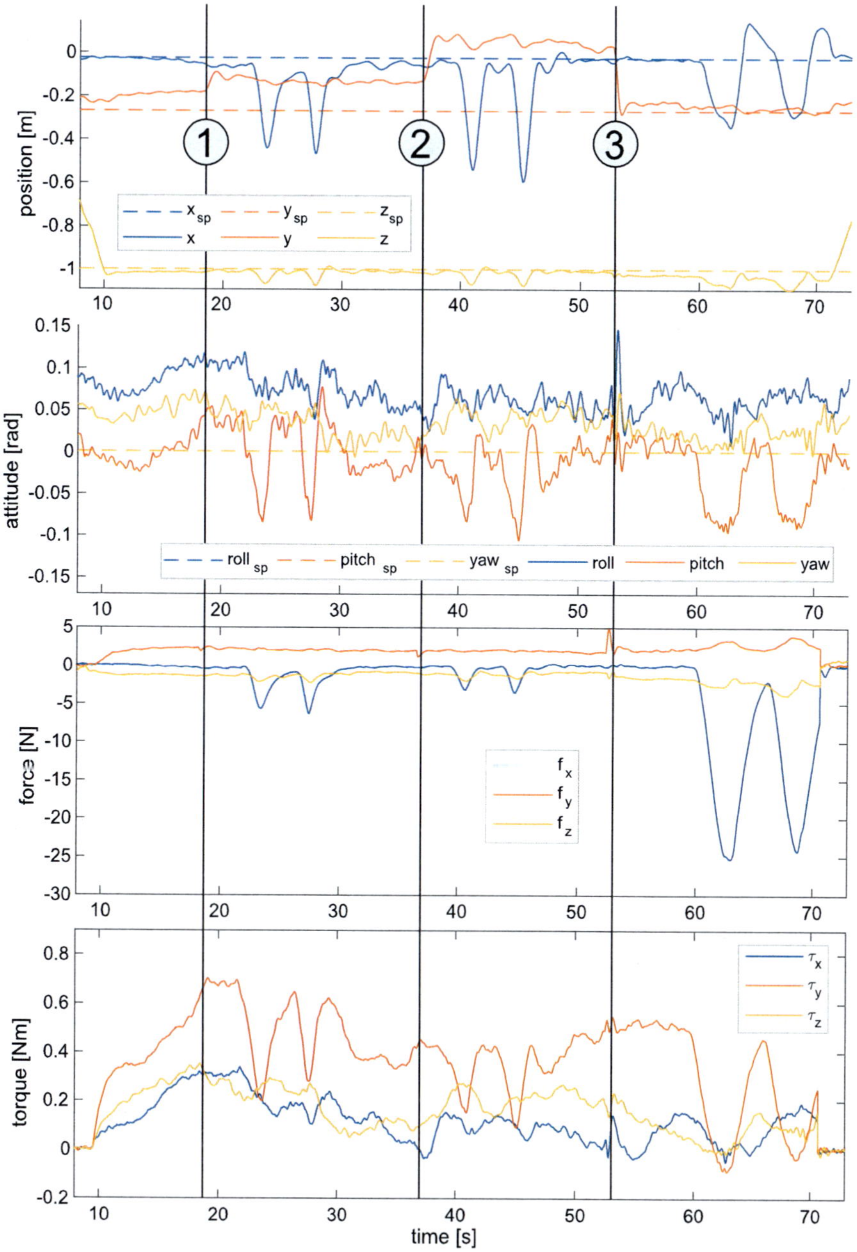

Fig. 6.23 Pose tracking and wrench estimation for rope pulling at the end-effector tip along the $\boldsymbol{e}_z^E$ axis in free flight

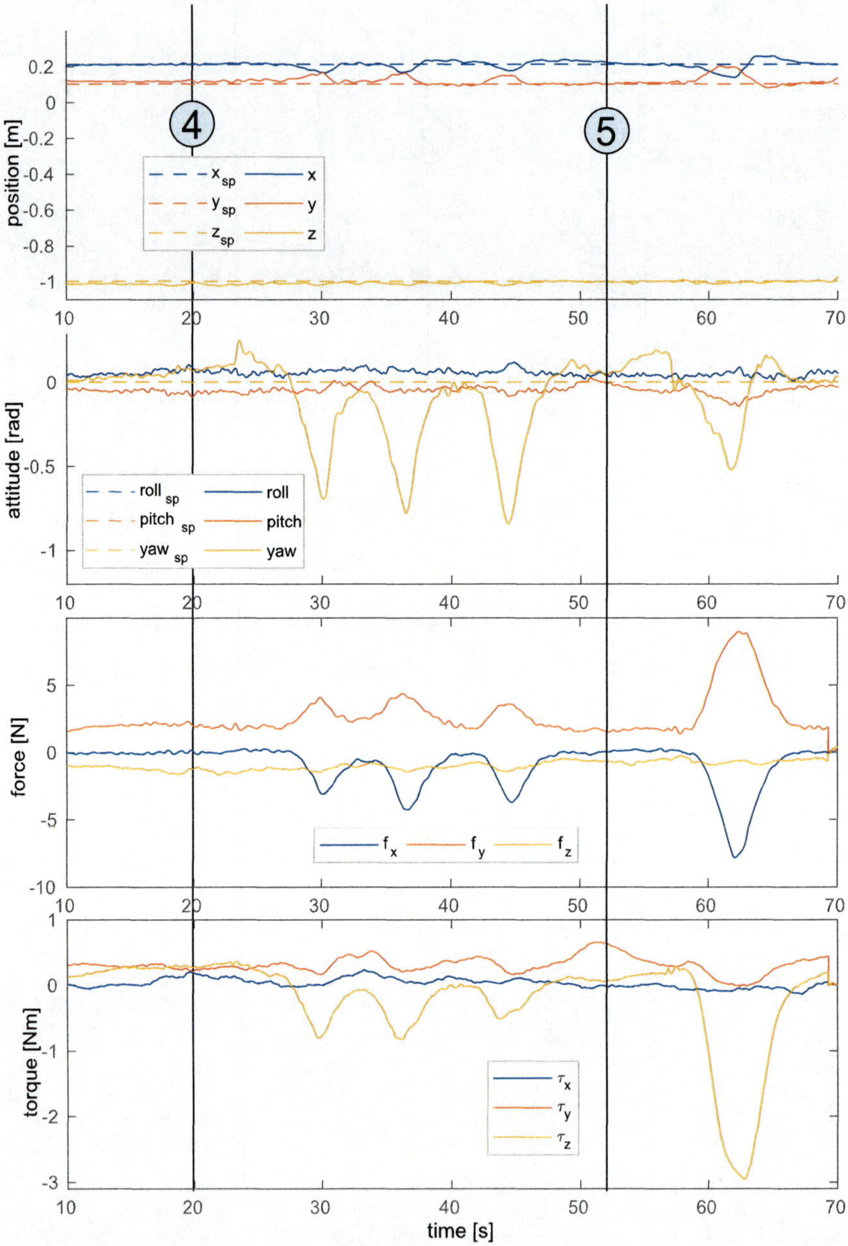

Fig. 6.24 Pose tracking and wrench estimation for rope pulling at the end-effector tip orthogonal to the $\boldsymbol{e}_z^E$ axis in free flight

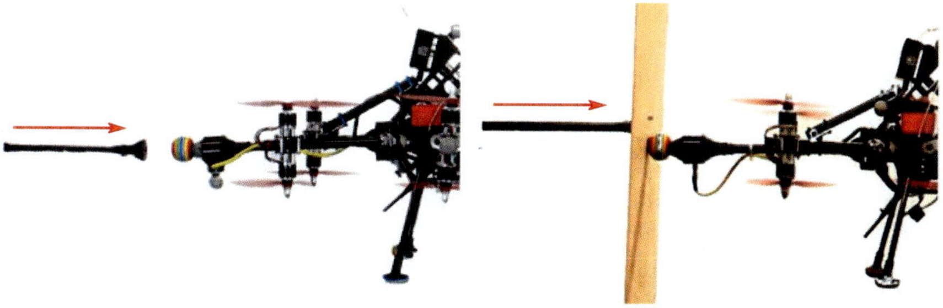

Fig. 6.25 Pushing with a stick (left) and a wooden panel (right). Direction of force is indicated by the red arrow

in free flight. For these tests, the system is again commanded to hold a reference altitude, and is installed with a forward-facing end-effector along the $\boldsymbol{e}_x^B$ axis. A disturbance is introduced by pushing the system from the end-effector point, applying force along the $\boldsymbol{e}_x^B$ axis. To simulate a free-flight 'unseen' disturbance, we push the system with a thin carbon fibre stick, and to simulate a contact surface, we mount a small wooden wall to the end of the stick which can be sensed by the TOF camera. A visual of these scenarios is shown in Fig. 6.25. Interaction force is measured by a force-torque sensor mounted below the end-effector for evaluation purposes, but is not used by the controller. The tests are performed with Proto^{V2} in the small flight arena using Vicon/IMU state estimation. Disturbances are introduced by hand, attempting to maintain the same motion of the stick or wall across all tests.

Table 6.12 provides an overview of test method and parameters for different disturbance trials, as well as the effective spring constant which is empirically calculated from filtered force and position error data along the contact axis. Tests 1–3 apply a disturbance to the system with a stick that cannot be observed by the distance sensor, and evaluate the system performance with axis-equivalent impedance control for different values of $m_v \in [1, 3, 5]$. Tests 4–8 apply a disturbance to the system with a moving wall, which the distance sensor can detect. Test 4 has the same parameters of test 1, to control the effect of a stick compared to a wall push. Tests 5–8 evaluate the effect of different values for contact compliance, with $m_{v,\text{wall}} \in [1, 0.5, 0.25, 0.1]$. Each of these tests is performed for fixed and variable axis-selective impedance control, where the latter are indicated with *. For wall-push tests using variable axis-selective impedance control, the virtual mass along the contact axis varies in the range $[m_{v,\text{free}}, m_{v,\text{wall}}]$ based on the observed distance, according to Eq. 4.54.

Figures 6.26a, 6.27a Measured and estimated forces for tests 1–3 are shown in Fig. 6.26a, with tests 4–8 shown in Fig. 6.27a (tests 5*–8* in the lower plot). Three disturbances are applied for each test configuration. Measured forces are smoothed by a forward-backward 5th order low-pass Butterworth filter with a 2 Hz cut-off frequency to capture the major trend in force and position error data without the influence of high-frequency vibrations and contact oscillations. Smoothed forces are

Table 6.12 Free flight disturbance test configurations, virtual mass parameters and resulting empirical spring constants with (variable) axis-selective impedance control. Virtual mass and inertia are multipliers for the system values, and the subscript 'v, wall' indicates the end-effector contact direction. The resulting the virtual mass matrix is constructed as $\mathrm{diag}([m_{v,\mathrm{wall}}\ m_{v,\mathrm{free}}\ m_{v,\mathrm{free}}\ \mathcal{I}_{v,\mathrm{free}}\ \mathcal{I}_{v,\mathrm{free}}\ \mathcal{I}_{v,\mathrm{free}}]^\top)$, since the contact surface normal aligns with the $\boldsymbol{e}_x^B$ axis. Spring constants, k, are estimated by a linear regression fit of the measured force and position error data shown in Figs. 6.26b, 6.28, along the body frame contact axis, where σ_f is the standard deviation of the force prediction error and b is the force axis intercept. *Indicated tests are performed with **variable** impedance along the contact axis

Test	Method	$m_{v,\mathrm{wall}}$	$m_{v,\mathrm{free}}/\mathcal{I}_{v,\mathrm{free}}$	k (N m^{-})	σ_f (N)	b (N)
Impedance control (all axes equivalent)						
1	Stick	–	1.0	137.36	2.79	6.05
2	Stick	–	3.0	165.50	3.89	8.67
3	Stick	–	5.0	164.22	4.24	9.53
Axis-selective impedance control						
4	Wall	1.0	1.0	113.43	1.24	10.31
5	Wall	1.0	5.0	126.53	1.18	10.88
6	Wall	0.5	5.0	82.08	2.20	12.23
7	Wall	0.25	5.0	52.58	2.21	9.96
8	Wall	0.1	5.0	12.29	2.95	8.92
***Variable** axis-selective impedance control*						
5*	Wall	(1.0)	5.0	120.06	1.88	9.32
6*	Wall	(0.5)	5.0	83.28	2.51	9.48
7*	Wall	(0.25)	5.0	57.29	3.44	8.38
8*	Wall	(0.1)	5.0	−0.48	2.73	8.93

plotted over raw measurements in the figures. We can see the effect of smoothing particularly clearly in Fig. 6.27b.

To evaluate the system's resistance or compliance to disturbance forces for different trials, we plot filtered interaction force against position error along the contact axis, as shown for tests 1–3 in Fig. 6.26b and 5-8(*) in Fig. 6.28. Hysteresis in the observed spring constant is apparent in the stiffness plot for the stick push with high virtual mass. This can be attributed to the relatively slow response of the momentum-based estimator with gain $\boldsymbol{K}_I = 1.0\mathbb{1}_6$ as well as the slow actuation bandwidth of the system. As a result, we only consider the region of rising force to determine the spring constant during disturbances.

For all plots, the low integral gain of the wrench estimator results in significantly reduced estimated force peaks compared to filtered ground truth measurements from the force sensor. The wrench estimator's ability to accurately estimate the contact force during transients has a strong effect on the impedance controller's behaviour in the presence of external impulses. In the case of our tilt-rotor system and chosen actuator allocation method, a low actuation bandwidth means that we can only counteract

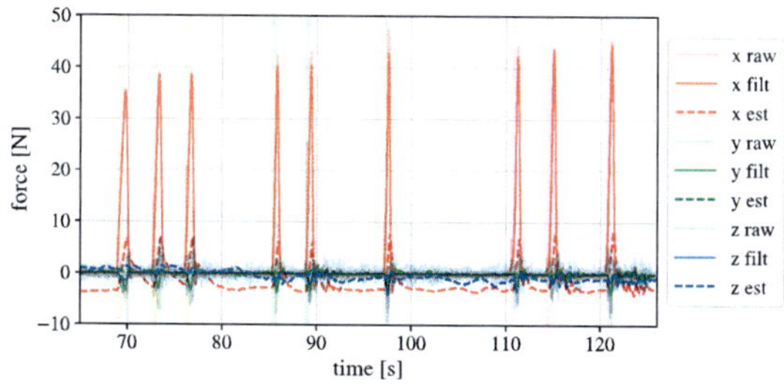

(a) Measured and estimated reaction forces in $\mathcal{F}_B$ for one stick disturbance experiment, where highlighted regions indicate the rising disturbance force. Data from these regions is used to calculate the effective spring constant, plotted in Fig. 6.26b.

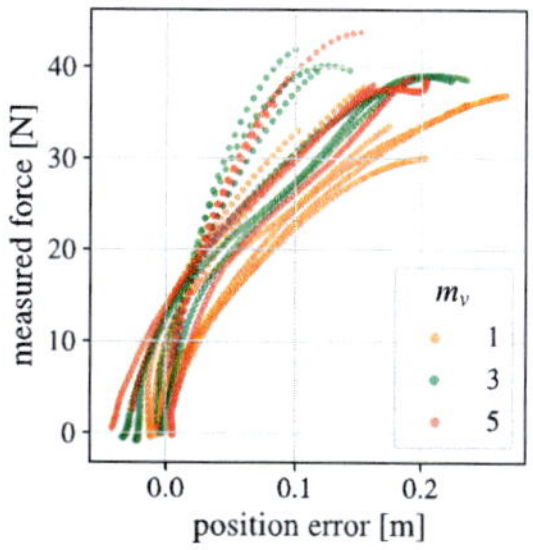

(b) Measured reaction force and position error are plotted for the e_x^B axis to show the effective spring constant during the rising phase of an introduced disturbance force.

Fig. 6.26 Free-flight trials where the system is disturbed by a stick. Disturbances are rejected in all directions using impedance control with a virtual mass that is 1, 3, or 5 times the system mass and inertia. Three disturbances are applied for each virtual mass setting, and the experiment is conducted twice. Measured forces are smoothed by a forward-backward 5th order low-pass Butterworth filter with a 2 Hz cut-off frequency

impulses below a certain frequency, and the low-pass nature of the momentum-based estimator improves flight stability.

For tests 1–3, for similar stick-push disturbances, we see the measured peak force grow with higher virtual mass. This reflects the intended behaviour of our system, where higher virtual mass rejects estimated disturbances more strongly. Observation of the system in flight led to the selection of $m_{v,\text{free}} = 5$ for the following experiments due to more consistent response, despite the similar performance to $m_{v,\text{free}} = 3$ in terms of spring constant. We attribute the minimal difference in spring constant between $m_{v,\text{free}} = 3$ and 5 to the actuation delay in response to an impulse. For a short disturbance, the entire region is within the initial response transient, limiting the disturbance-rejecting force.

We compare the spring constant result of tests 1 and 4 to confirm that the empirical spring constants are similar in magnitude for disturbances introduces by a stick and a wall, where the values of 137.36 and 113.43 N/m are within a range small enough to confirm this assumption. Comparing tests 4, 5 and 5*, similar spring constant values indicate that virtual mass properties along the $\boldsymbol{e}_x^B$ axis are effectively decoupled, regardless of the virtual mass assigned to the remaining axes.

We consider Fig. 6.27a and the magnified time range in Fig. 6.27b to compare estimated and measured force data for fixed and variable axis-selective impedance control in tests 5(*)–8(*). Across applied disturbances for the lower plot with constant compliance along the end-effector axis, the momentum-based force estimate deviates more strongly in free flight from the expected zero value with increasing compliance. This deviant force estimate reflects the increasing effect of model error in the direction of contact, since model error is lumped with the contact force for momentum-based wrench estimation, also corresponding with a larger position error along $\boldsymbol{e}_x^B$ as the

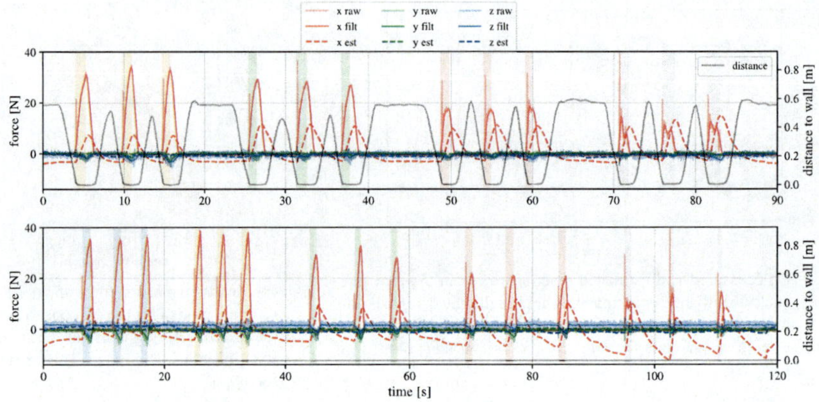

(a) Wall push disturbance trials for variable (upper) and non-variable (lower) axis-selective impedance control, corresponding with tests listed in Table 6.12. Highlighed areas indicate periods of rising contact force, and values in these regions are used for the calculation of effective spring constants.

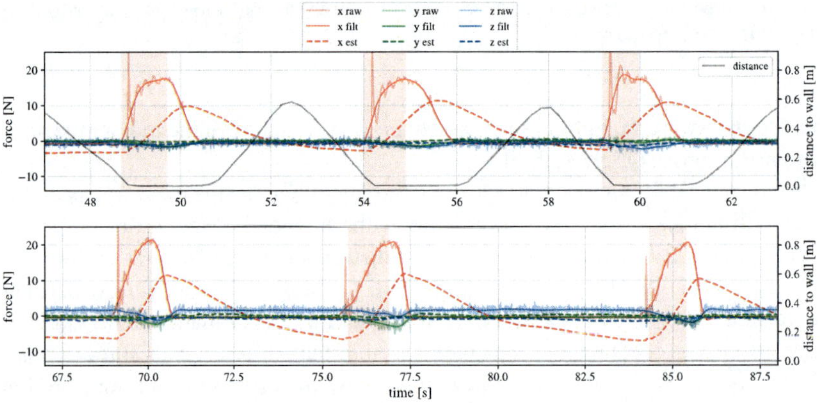

(b) Magnified plots of three wall-push disturbances with $m_{v,\text{wall}} = 0.25$ from plots in Fig. 6.27a. Results for variable (upper) and non-variable (lower) axis-selective impedance control differ notably in momentum-based force estimates between pushes.

Fig. 6.27 Flight experiments with disturbances introduced by a moving wall along the $\boldsymbol{e}_x^B$ axis, with reaction forces expressed in $\mathcal{F}_B$

system exhibits compliance to the error. Variable axis-selective impedance control addresses this issue in free flight. We observe the distance sensor measurement fall below $d_{\min} = 0.2\,m$ in the contact region, and retract during free flight, resulting in strong disturbance rejection along the contact axis between wall pushes. This result indicates that varying impedance along the contact axis in response to a distance measurement is effective in reducing free flight compliance to model error.

We further evaluate the difference in computed spring constants between fixed and variable axis-selective impedance control for the selection of $m_{v,\text{wall}}$, observing the relationship plotted in Fig. 6.28. In general, we clearly see the expected reduction in spring constant with lower $m_{v,\text{wall}}$ for both fixed and variable compliance, as

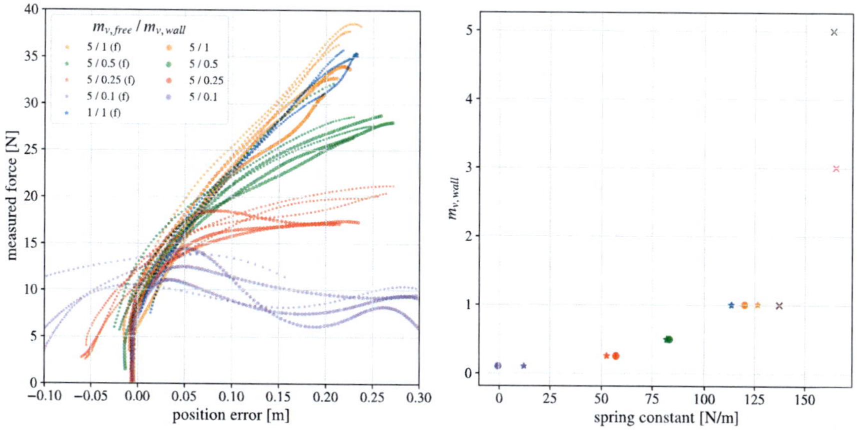

Fig. 6.28 **Left:** Position error is plotted against the filtered force measurements, representing the spring constant for free-flight disturbances with (variable) axis-selective impedance control. Star markers, indicated with in the legend with (f) represent trials where $m_{v,\text{wall}}$ is fixed to be compliant. Circle markers indicate variable $m_{v,\text{wall}}$, returning to $m_{v,\text{free}}$ with increasing distance form an observed surface along the contact axis. **Right:** The trend for spring constant vs. $m_{v,\text{wall}}$ is shown, including all values from Table 6.12

quantified in Table 6.12. A negative spring constant for test 8*, as well as the steeper initial slope for all trials is the combined result of the sudden initial impulse and delay in actuation response. This effect results in a similar force-axis intercept value for all trials, which supports the similarity of all wall-push motions. We further note that where interaction starts, fixed compliance on the end-effector axis has allowed a large accumulation of position error due to compliance to model error. Variable compliance on the other hand brings position error close to zero when the wall is not detected close to the contact point, and is still able to exhibit desired compliant behaviour during the disturbance. This observation further demonstrates the advantage of variable axis-selective impedance control for simultaneously managing model error and contact disturbances.

In the right plot of Table 6.12, we see the relationship of all spring constants from Table 6.12 plotted against virtual mass values, following a trend for which the linear range is limited by the actuation bandwidth and the delay in momentum-based force estimation.

6.3.3 *Force Tracking Accuracy*

To evaluate the accuracy of force tracking with direct interaction force control (as a component of hybrid force-impedance control), we design a physical interaction trajectory containing both a pose and a force reference. The position reference is set on the known surface of a rigid vertical wall, and the attitude reference sets the $\boldsymbol{e}_z^E$

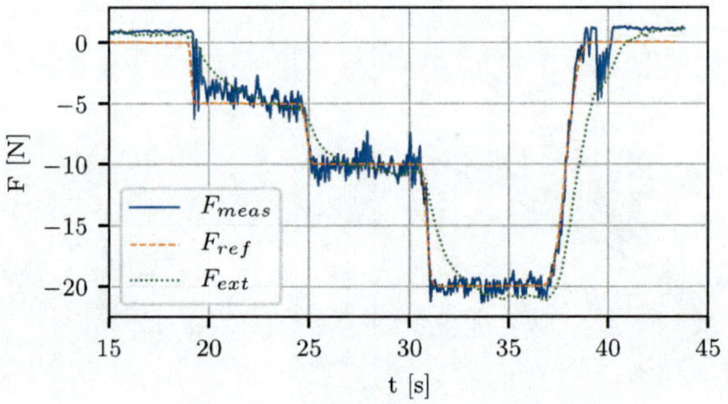

Fig. 6.29 Tracking of 3 different force references. The dashed orange line represents the force reference in the world frame, the solid blue line is measured force at the end-effector in the contact direction, filtered as described in Sect. 6.3.1. The dotted green line shows the momentum-based force estimate

axis orthogonal to the surface plane. The force trajectory is oriented opposite to the surface plane normal, and changes between 5, 10, and 20 N. This test is performed with Proto^{V2} in the large flight arena using Vicon/IMU state estimation.

Figure 6.29 shows the tracking performance of the force controller. The reference force is tracked consistently, even through fast changes of the set point. Though it is not used during the force control phase, we see that the momentum-based force estimate adapts more slowly to the change of force but converges to similar values.

For tuning of the force control PID, we found that both the feed-forward term and the integral gain are the most essential parts of the force controller. Proportional gain is set to a small value and differential gain to zero, due its destabilizing effect. Increasing the proportional gain leads to higher frequency changes in the resulting contact forces, while not significantly improving the response time to reference changes. The plots shown use the force control PID gains listed in Table 6.9.

Despite action taken to avoid force sensor drift, Fig. 6.29 indicates an erroneously measured offset. As discussed in Sect. 6.3.1, cooling effects of the propellers can be significant, and a slowly changing bias that was not reset resulted in an offset. Regardless of the offset, the resulting performance demonstrates an ability to quickly respond to measurements from an integrated force sensor. The remaining error can be corrected with further investigations into sensor calibration.

Robustness to Planning Errors

Additionally, we test the robustness of the proposed hybrid force-impedance control (described in Sect. 4.2.3) to small errors in the planned trajectory by performing a surface inspection task with multiple contacts with reference forces of 5 N. In a first experiment, we set the position reference 4 cm in front of the true surface position, and in a second trial we set it 1 cm behind the true position. For both trials, the system is close enough to the surface that the confidence factor, $\lambda_f = 1$, such that direct force control is fully active.

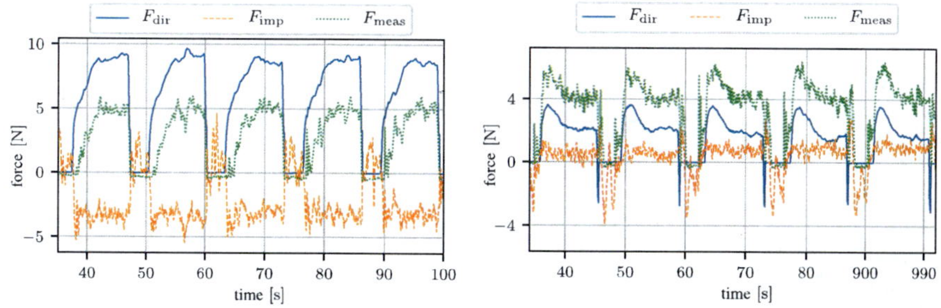

(a) Position reference 4 cm in front of the wall.

(b) Position reference 1 cm behind the wall.

Fig. 6.30 Hybrid force-impedance control experiments showing robustness to position reference in front of or behind the contact surface. A 5 N contact force is commanded for 10 s intervals. Plots compare measured interaction force (F_{meas}) with direct force control (F_{dir}) and impedance control (F_{imp}) components of the commanded force in the direction of contact

Plots of the body force command in the contact direction for the two trials are shown respectively in Fig. 6.30a and b, also depicting the interplay of force and impedance control components as described in Eq. 4.68, and shown as a block diagram in Fig. 4.14. The figures illustrate that for each trial, the measured force settles at the reference force after a short response time. The impedance control component holds a constant offset term based on the position error when the end-effector is against the contact surface. The direct force control term compensates for this error, driving the force error to zero primarily with the integral term. The feed-forward force control term initially causes an undershoot in force (Fig. 6.30a) where the position set point is away from the wall, and an overshoot in force (Fig. 6.30b) where the set point is into the wall. The experiments show that the direct force control command $\boldsymbol{f}_{dir}$ compensates the impedance control command $\boldsymbol{f}_{imp}$ to achieve the desired contact force.

When planner error is larger, we enter the domain where $\lambda_f < 1$, the transition zone of attempting direct interaction force control. To evaluate the robustness and behaviour of hybrid force-impedance control in the case of large planner error, we specify a desired contact force of 10 N at three different distances from the contact surface:

(a) approximately on the surface
(b) 0.25 m away from the surface
(c) 0.5 m away from the surface

Figure 6.31 shows the resulting behaviour for the three scenarios. At point (a), the platform is close enough to the wall such that $\lambda_f = 1$ and direct force control is fully enabled. The lower plot shows force tracking results, where the desired 10 N magnitude is reached. This case exhibits the desired interaction control behaviour if the reference is well positioned close to the surface. At point (b), the perceived end-effector distance d_E in combination with the end-effector position error $\boldsymbol{e}_t$ places the system in the force confidence transition zone based on our chosen interaction

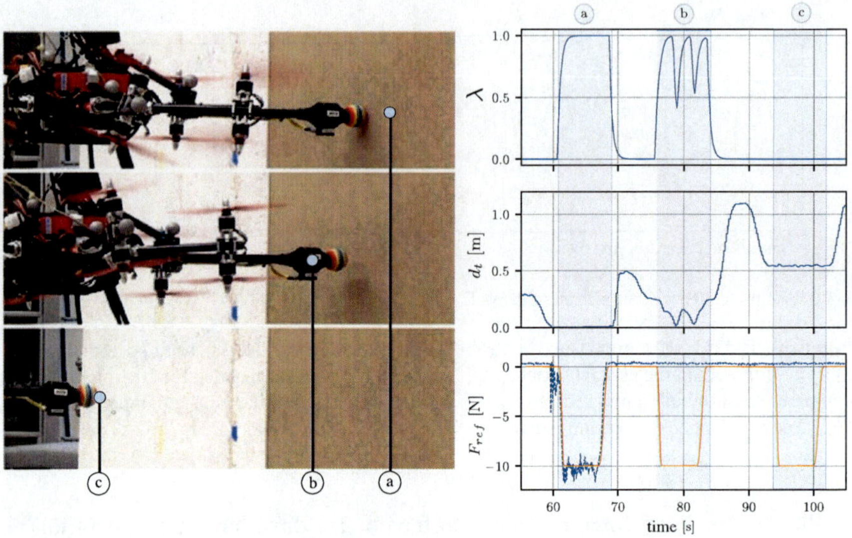

Fig. 6.31 Confidence factor λ, wall distance and force reference show behaviour for a set point **a** behind the wall, **b** in front of the wall by 0.25 m, and **c** >0.5 m away from the wall

control parameters. A short increase in λ is seen as the system attempts to engage in force control, before the end-effector error grows sufficiently to reduce the confidence factor, and the resetting force of the impedance controller guides the system back to the pose reference. No contact is made in this case. At point (c), the end-effector distance d_E is larger than the maximum selected end-effector distance $d_{\max}$, resulting in a confidence factor of $\lambda_f = 0$, and direct force control is therefore not enabled.

In all cases, the system responds to planner error in a stable way, and is able to continue executing a compromise of the combined state and force trajectory. The behaviour of each case depends on chosen values for $d_{\min}$, $d_{\max}$, $e_{\min}$ and $e_{\max}$, and how they generate λ as described in Eq. 4.61. Case (a) and (c) are desired effects when the system is close enough that establishing contact is more important than position tracking, and when the system is far enough from a surface that force tracking should not even be attempted. In case (b), which is visualized in Fig. 4.13, the contact surface is detected but the position error is too large to track the end-effector tip at the surface. This case can be avoided by choosing $d_{\max} \leq e_{\min}$. Provided that the nominal tracking error remains small, the system remains in a position tracking region until the detected surface is within the tracking error margin.

This controller behaves well as long as the platform is able to effectively measure the distance to the surface. Several real-world scenarios can complicate this task, for instance if surfaces are transparent or reflective and therefore not 'seen', or if the environment is too dark or obscured with particulates, such that a surface is registered even when it isn't present. These cases motivate the need for multi-sensor integration, where the strengths of different sensor types can be combined to generate a better-informed confidence factor for interaction.

6.3.4 Push-and-Slide Aerial Drawing

In this experiment, we evaluate the ability of the system to maintain a normal orientation to a whiteboard, rejecting disturbances from friction forces when interacting, as well as the ability to accurately and repeatably draw a defined pattern on the surface. Both axis-selective impedance control and direct force control (in the hybrid force-impedance control framework) are evaluated. The whiteboard is positioned in a known location, and the end-effector is a whiteboard marker with no additional compliance.

Axis-selective impedance control experiments are performed in the large flying area with Proto^{V1}, and direct force control experiments are performed in the small flying arena with Proto^{V2}. Interaction control parameters are listed in Tables 6.5, 6.9. Both sets of experiments use an external motion capture system fused with measurements from the onboard IMU.

A: Axis-Selective Impedance Control

For **impedance control** interaction experiments, a trajectory traces a drawing with the end-effector point 10 cm behind the surface of the whiteboard. Virtual mass and inertia are set to 5 times that of the system in all directions, except for the $\boldsymbol{e}_z^E$ axis, where virtual mass is 0.25 times the system mass.

Tracking results for position and orientation in the top two plots of Fig. 6.32 show ground truth measurements from the motion capture system of two trials drawing the same shape on a whiteboard, compared to the reference trajectory set point, marked with subscript sp. In the time interval between (a) and (b), the end-effector is in contact with the whiteboard, maintaining a consistent estimated force while completing a trajectory.

The system demonstrates its ability to handle transitions in and out of contact with good stability, and trace a contact trajectory without significant tracking error on the surface plane. The system demonstrates a good ability to reject torque and lateral force disturbances caused by surface friction while maintaining a consistent contact force against the wall, as shown in the lower two plots of Fig. 6.32. The contact force of about 2 N is low enough to avoid static friction.

Offsets in $\boldsymbol{e}_x^B$-force and $\boldsymbol{e}_y^B$- and $\boldsymbol{e}_z^B$-torque in free flight—as well as a small bias in the attitude tracking of the system—are the result of an unaccounted-for offset of the system's center of mass. This result demonstrates that the proposed impedance control can compensate well for model errors, maintaining attitude error within 0.07 rad.

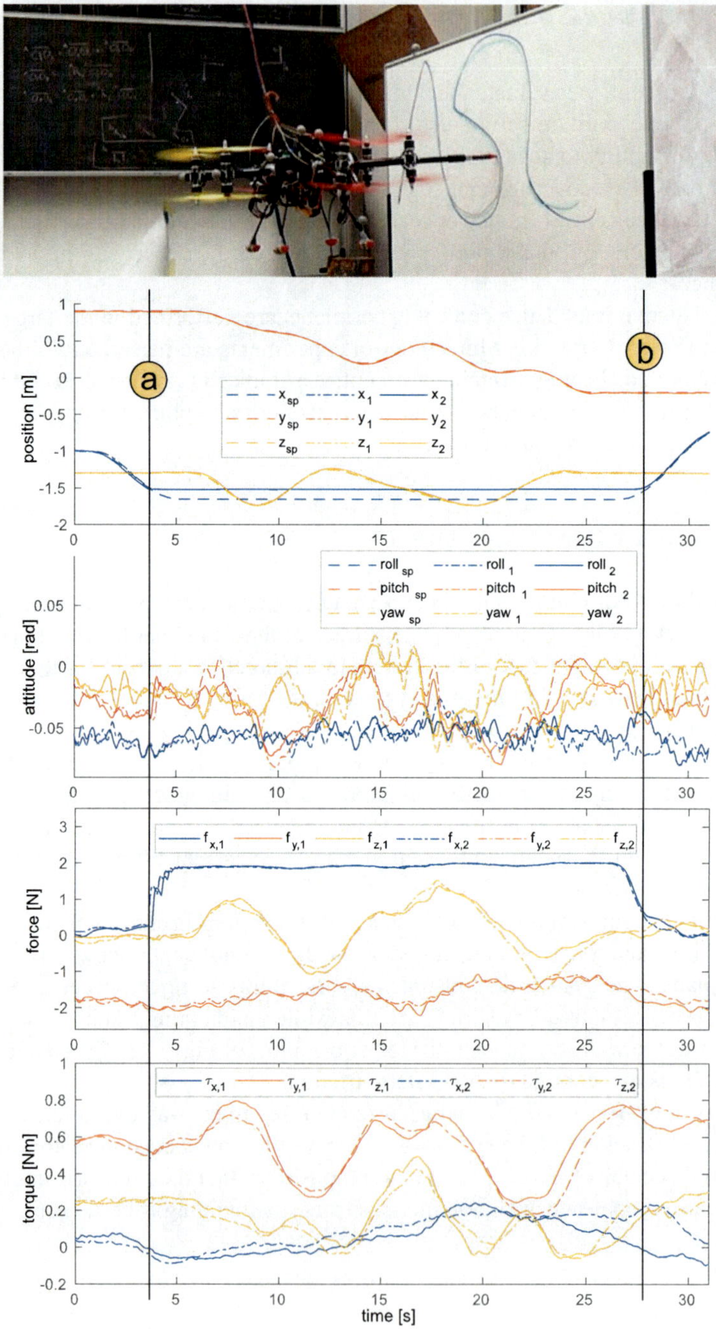

Fig. 6.32 Pose tracking and momentum-based wrench estimates for push-and-slide experiments. A shape is drawn on a whiteboard aligned with the $\boldsymbol{e}_x^W$ plane in two separate trials (subscripts 1 and 2). At time **a**, the system contacts the wall, and at **b**, resumes free flight

Fig. 6.33 Left: an image from experimental trials. Right: the resulting whiteboard drawing from 9 subsequent force control trials (3 each of 1, 3 and 5 N reference forces)

Table 6.13 Base tracking error statistics for push-and-slide aerial drawing experiments. The projected drawing plane mean error and total angular error compare overall performance between experiments. Position errors are expressed in meters, and attitude errors in radians

		x	y	z	x-z norm	Roll	Pitch	Yaw	Angle
Imped.	Mean	−0.000	**0.033**	−0.008	**0.017**	−0.003	0.057	0.008	**0.060**
	std.	0.006	**0.010**	0.019	0.012	0.005	0.009	0.018	0.010
	RMS	**0.006**	0.035	**0.020**	0.021	**0.006**	**0.058**	**0.020**	0.061
5 N	Mean	−0.001	**−0.017**	−0.009	**0.015**	−0.000	0.058	0.013	**0.066**
	std.	0.008	**0.010**	0.012	0.008	0.006	0.014	0.029	0.015
	RMS	**0.008**	0.020	**0.015**	0.017	**0.006**	**0.059**	**0.032**	0.068
3 N	Mean	−0.001	**−0.018**	−0.008	**0.016**	0.002	0.059	0.011	**0.064**
	std.	0.007	**0.010**	0.015	0.010	0.005	0.012	0.022	0.013
	RMS	**0.007**	0.021	**0.017**	0.019	**0.006**	**0.060**	**0.025**	0.066
1 N	Mean	−0.000	**−0.017**	−0.008	**0.016**	0.001	0.053	0.006	**0.056**
	std.	0.008	**0.010**	0.016	0.012	0.006	0.008	0.014	0.008
	RMS	**0.008**	0.020	**0.018**	0.020	**0.006**	**0.054**	**0.015**	0.056

B: Direct Interaction Force Control

For **direct force control**, the trajectory traces a spline with the end-effector point on the surface of the whiteboard. The force trajectories smoothly ramp up to the desired force reference after the position reference has reached the known whiteboard surface location. We compare the tracking performance of three different force references of 1, 3, and 5 N, along with axis-selective impedance control as a baseline. Trials are performed three times with each set of parameters to allow for better statistical analysis. Figure 6.33 shows a photo taken during the experiment, and the resulting whiteboard drawing from all force control trials.

Table 6.14 End effector position and force tracking error statistics for push-and-slide aerial drawing experiments. The mean of projected drawing plane error and the interaction force error are highlighted for each experiment. Position errors are expressed in meters, and force errors in Newtons

		x (m)	z (m)	x-z norm	f_y (N)
Imped.	Mean	0.002	−0.013	**0.022**	–
	std.	0.007	0.022	0.015	–
	RMS	**0.007**	**0.025**	0.026	–
5 N	Mean	−0.000	−0.017	**0.026**	**−0.383**
	std.	0.009	0.022	0.014	0.752
	RMS	**0.009**	**0.028**	0.029	0.844
3 N	Mean	−0.002	−0.014	**0.024**	**−0.699**
	std.	0.008	0.023	0.015	0.973
	RMS	**0.009**	**0.027**	0.028	1.198
1 N	Mean	−0.000	−0.011	**0.019**	**−0.579**
	std.	0.010	0.018	0.013	0.917
	RMS	**0.010**	**0.021**	0.023	1.085

Tracking error statistics for the system base are shown in Table 6.13, while Table 6.14 presents end-effector error on the drawing plane and interaction force tracking error. Corresponding violin plots aid in interpreting the results, showing the error distributions for base and end-effector tracking in Fig. 6.34, and force tracking error distributions in Fig. 6.35.

Table 6.13 shows that the base reference position is tracked similarly well for the 4 different cases, with a projected mean error of 1.5–1.7 cm. Comparing RMSE values for $\boldsymbol{e}_x^W$ and $\boldsymbol{e}_z^W$, we attribute larger error on the $\boldsymbol{e}_z^W$ axis due to inertial model error amplified by gravitational effects. Angular errors for the 4 cases have similar mean values (0.056–0.066 rad), though larger mean and standard deviation are seen for force control trials with higher commanded tracking forces (standard deviations of {0.015, 0.013, 0.008} rad for {5, 3, 1}N respectively). This effect can be explained by increased contact force leading to higher lateral friction at the end-effector point, causing a moment about the pitch and yaw axes of the floating base.

Position error for end-effector tracking is only evaluated in unconstrained directions (on the whiteboard surface, i.e. the $\boldsymbol{e}_x^W$-$\boldsymbol{e}_z^W$ plane). To fairly compare base and end-effector tracking error, we present the same projected error for the base. The projected end-effector position exhibits a larger tracking error than the base (1.9–2.6 cm vs. 1.5–1.7 cm mean error), as seen in Fig. 6.34 and numerically presented in Table 6.14. Since the controller is explicitly tracking a base reference and attitude that correspond with the desired end-effector trajectory, our PID feedback law acts at the body center, and as a result any base orientation error magnifies position error at the end-effector. Despite using high stiffness impedance control parameters for attitude and lateral position, unmodeled effects such as surface friction at the whiteboard, as well as a larger end-effector mass due to the force sensor, prevent more accurate end-

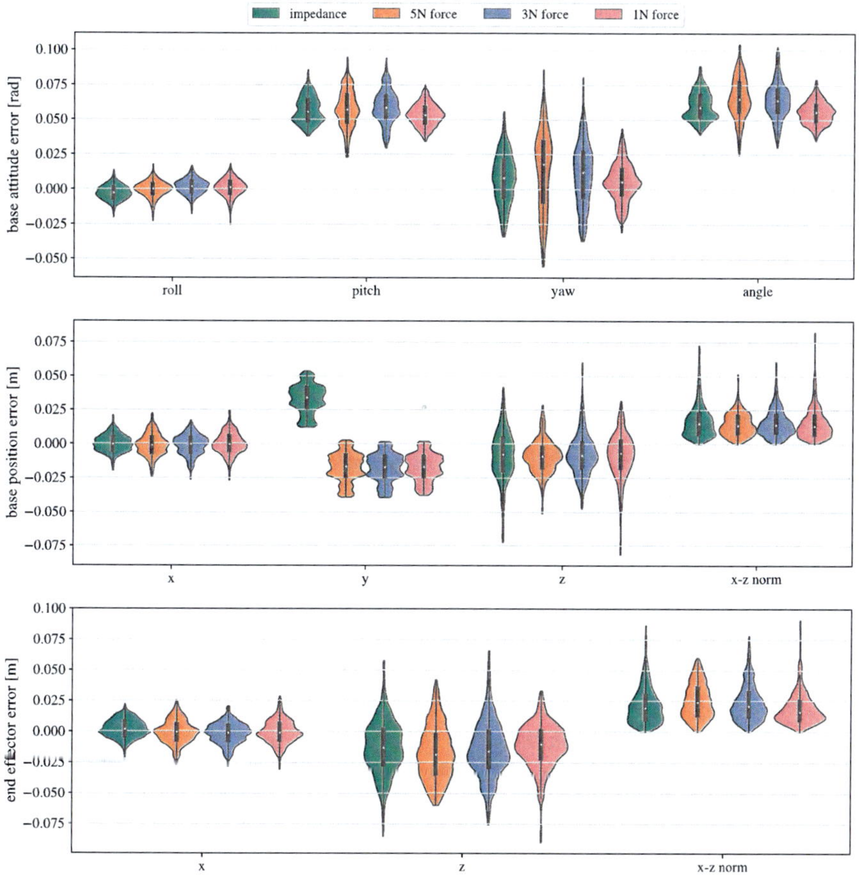

Fig. 6.34 Error distributions for (top) the base attitude, (middle) the base position and (bottom) the end-effector position are shown in the form of violin plots. Data is collected from the in-contact drawing phases of three trials for each control scenario. The x-z norm is used for position errors to for comparison of the projected error onto the trajectory plane at the base and end-effector points. Error for the end-effector y-axis is not shown since it is physically constrained by the whiteboard surface. Attitude is equivalent for the base and end-effector since they are rigidly connected

effector tracking. This issue can be resolved by explicitly tracking the end-effector point, which then requires the additional incorporation of contact constraints in the controller.

Base $\boldsymbol{e}_y^W$ position error is also shown in plots for comparison, where all force control trials show a mean error around $-1.7\,\text{cm}$, with $1\,\text{cm}$ standard deviation. This mean value indicates that the trajectory is on average located in front of the whiteboard surface by this amount, where the standard deviation can be attributed to slight misalignment with the $\boldsymbol{e}_x^W$-$\boldsymbol{e}_z^W$ plane, small deflections of the whiteboard during interaction, and deformation of the marker tip. For impedance control, the mean $\boldsymbol{e}_y^W$

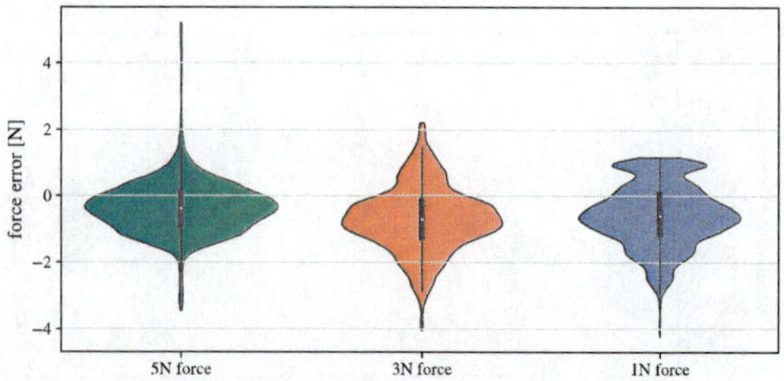

Fig. 6.35 Force tracking error distributions are presented for commanded force scenarios of 1, 3, and 5 N. Data is collected from the in-contact drawing phases of three trials for each force control scenario

error of 3.3 cm corresponds to the trajectory's submersion behind the nominal surface plane to generate contact force.

Force tracking results in Table 6.14 and Fig. 6.35 show similar error statistics for all three reference force magnitudes. Negative mean values for all force control trials indicate consistent undershooting of the desired force, which can be attributed to rise time to reach the desired force and drifting force sensor biases which cause a reference offset. For the 1 N tracking case, the distribution has a second peak at 1 N error which indicates the contact transition zone. Consistent with the standard deviation of force tracking error across all trials, tracking a force this low results in periodic loss of contact.

To further investigate end-effector tracking and interaction force over the trajectory, we overlay informative data on the projected drawing plane. Figure 6.36 plots the base position projected on the e_x^W-e_z^W plane, where data points are coloured by the measured interaction force. Three overlaid trials for each of four experimental

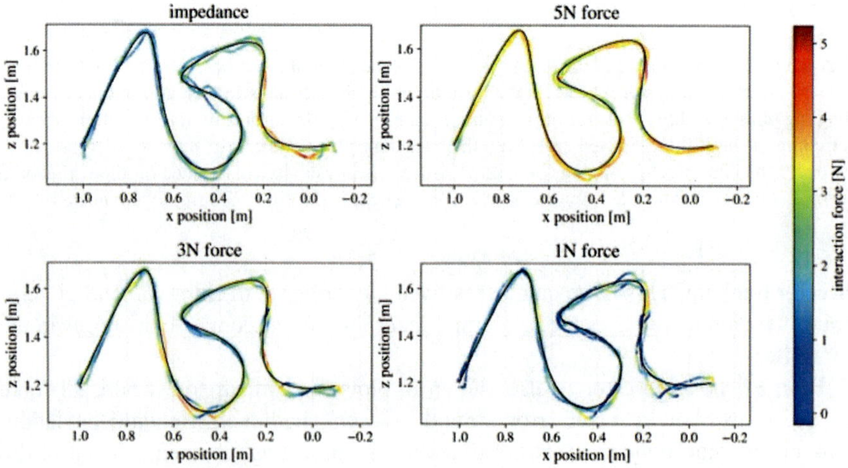

Fig. 6.36 The system base position is plotted on the x-z plane in $\mathcal{F}_W$ for different control scenarios, each consisting of 3 trials. Colours indicate the interaction forces measured by the on-board force sensor in the surface normal direction, and the trajectory is indicated with a black line

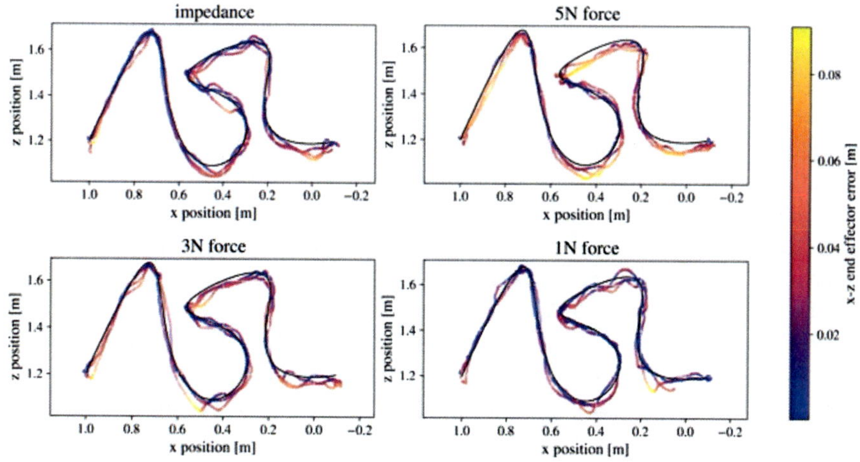

Fig. 6.37 The end-effector position is plotted on the x-z plane in $\mathcal{F}_W$ for different control scenarios, corresponding with the trials in Fig. 6.36. Colours indicate the end-effector position error norm projected onto the x-z plane, and the trajectory is indicated with a black line

conditions are plotted in a grid for comparison, where the reference trajectory is indicated in each figure by a black line. For force control trials of 1, 3, 5 N, coloured trajectory data fluctuates about the reference. For all trials, certain regions of the trajectory consistently correspond with lower (e.g. $x, z = 0.4, 1.4$) and higher (e.g. $x, z = 0.2, 1.6$) forces. These regions appear to be related to the trajectory direction and curvature, which affect the moment caused by friction forces on the system base. Near-black data points in the 1 N force case which trace straight lines (e.g. $x, z = 0.4, 1.1$) indicate loss of contact. For the impedance control case, we see lower forces at the left of the trajectory transitioning to higher forces with decreasing x position. This trend indicates a slight misalignment in the $\boldsymbol{e}_x^W$-$\boldsymbol{e}_z^W$ and whiteboard planes, such that the reference position depth behind the whiteboard changes slightly on the horizontal axis. Force control has an advantage in this respect, that the reference force is tracked explicitly. This force consistency could otherwise be solved with a live distance measurement to the surface, and re-projection of the trajectory to maintain a constant depth reference.

Figure 6.37 presents similarly projected data, plotting the end-effector position on the whiteboard surface, where data points are coloured by the projected position error norm. The end-effector trajectory visibly deviates from the desired trajectory more than that of the base, as anticipated in our error analysis. This effect is more prominent in the right half of the drawing. Considering the force control trials, we clearly see larger end-effector error with increasing force reference and particularly along the $\boldsymbol{e}_z^W$, as seen in the error data. This plot allows us to see more clearly that while tracking a 1 N reference has a tendency to oscillate about the reference in horizontal motion, tracking a 5 N reference exhibits a consistent trajectory offset in horizontal motion due to the effects of increased surface friction.

The results of aerial writing experiments allow us to draw a few important conclusions about this type of physical interaction task. Lateral friction forces pose a challenge to trajectory tracking, forcing a trade-off between contact force and end-effector position tracking performance. One major reason for this is the implicit time constraints in each trajectory. Relaxing the trajectory time constraint to a path-following problem in cases where exact timing is not critical will allow for improved performance in both end-effector position and contact force. Tracking a reference at the base will only produce equivalent results for the end-effector if there were no orientation error. Therefore improved tracking performance will require formulating the PID control feedback term around end-effector error, and integrating contact constraints at this point.

For any interaction controller, a higher control wrench bandwidth will improve overall system performance, allowing for fast and accurate disturbance rejection of friction forces and wind gusts. Model error also significantly affects the control wrench accuracy, particularly at lower bandwidths. A significant increase in hover error is present in experiment series **B** with Proto^{V2} compared to **A** with Proto^{V1}. Experiments with Proto^{V1} were performed on the system in a relatively new state, with good system model accuracy. With Proto^{V2}, the system had already sustained numerous crashes and repairs, with the geared tilt-rotor mechanisms exhibiting increased backlash over time. The resulting degraded system model accuracy is evident when comparing the quality of impedance control drawing results in Figs. 6.32 and 6.33.

For direct force control, a combination of actuation bandwidth and force reference signal is key to reducing force tracking error. Since force error directly contributes to the control wrench, the correction frequencies should be achievable by the control wrench such that measured force vibrations do not produce a positive feedback response. A smooth force signal with minimal delay in combination with high bandwidth base wrench control will allow us to raise force control gains and improve the force tracking response. Impedance control performs well for push and slide tasks, and could also be used to track a desired reference force if the correlation between trajectory depth behind a surface and the resulting contact force is well established. The distance sensor already integrated for hybrid force-impedance control can be used to maintain the reference at a depth corresponding to the desired force. This approach would remove the need for force sensor signal processing, but relies on a very good system model, such that model error has a negligible contribution to the estimated contact wrench.

In summary, direct force control is a good solution when we have a high base wrench actuation bandwidth and smooth force feedback signal with low delay, despite large system model error and an uncertain environment. Impedance control is a good solution when we have a very good system and environment model, despite lower control bandwidth and insufficient force sensor data.

Fig. 6.38 Image from experimental trials for (left) rolling in contact on an archway with impedance control and (right) random normal point contacts on an undulating wooden surface with direct force control

6.3.5 *Interaction with Inclined Surfaces*

The physical interaction experiments presented in Sects. 6.3.1, 6.3.2, 6.3.3, 6.3.4 have been limited to planar vertical surfaces. Although these present a good baseline for the system performance, they do not take advantage of the platform's omnidirectionality.

This section explores some interaction cases on curved surfaces with changing inclination relative to gravity. In the first experiment, we perform sustained contact while rolling on the underside of an archway using axis-selective impedance control with on-board VIO-based state estimation and local planning. In a second set of experiments, we perform force control at a random selection of contact points on an undulating surface, which is inclined both positively and negatively to the gravitational vector. A large number of point contacts allows for a comprehensive statistical analysis of the data. Images of the two experiments are shown in Fig. 6.38.

A: Rolling in Contact on an Arch

To evaluate the performance of impedance control for changing surface inclinations, we conduct a rolling-in-contact test for the system on the underside of an archway. In this trial, we use Proto^{V1} with fully on-board state estimation which fuses VIO with IMU measurements, which is also used for tracking analysis. The end-effector is oriented at a 30° rotation about $\boldsymbol{e}_y^B$, as seen in Fig. 5.12a. A wheel is installed at the end-effector to reduce friction forces during interaction, and is aligned in advance with the desired direction of motion during contact. The value of $m_{v,\text{wall}}$ for these experiments is raised from 0.25 to 0.5 to improve the reliability of contact.

Depth-servoing is used to maintain orientation to the local surface normal, as described in Sect. 6.1.3. An initial trajectory guides the system to a start point away

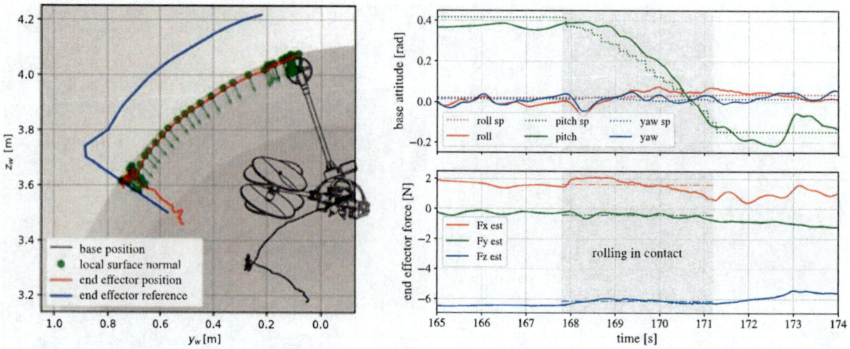

Fig. 6.39 **Left:** End effector position and reference are plotted on the $\boldsymbol{e}_y^W - \boldsymbol{e}_z^W$ plane, along with the base position while rolling in contact on an archway. Local surface points and normal vectors used for real-time trajectory updates are indicated with green arrows. **Right:** Attitude tracking plot for a changing pitch reference, where the end-effector reference moves during the shaded region. The lower plot shows the momentum-based force estimate in the flying base frame, with mean values over the shaded range indicates with a dash-dot horizontal line

from the surface, orienting the end-effector $\boldsymbol{e}_z^E$ axis to align with the locally observed arch surface normal. The end-effector is then brought in contact with the surface, and the reference is updated with live surface normal measurements sampled at 5 Hz to trace a path along the curved arch surface. The end-effector traverses a distance of 0.53 m on the surface during contact, with an average velocity of 0.17 m s^{-1}. Tests are performed with no prior information of the structure, using manual initial positioning and local sensing for the contact trajectory.

For this experiment, the major criteria for success is maintaining contact while rolling along the structure, indicated by the turning end-effector wheel in the corresponding video. This trajectory aligns the compliant contact axis with gravity, to evaluate impedance control performance in a worst-case model error scenario. For evaluation, we consider attitude tracking of the system and interpret the momentum-based force estimate.

As visualized in the left plot of Fig. 6.39, locally perceived surface points and normal vectors produce an end-effector reference that ensures constant contact while rolling along the arch, robust to state estimator drift. Update discretization of 5 Hz is visible in the pitch reference of the right plot, which is tracked with a slight delay and overshoot at the end of the rolling motion. Despite model error, the impedance controller generates sufficient force along the $\boldsymbol{e}_z^E$ axis to maintain contact. The lower right plot of Fig. 6.39 shows the momentum-based force estimate in $\mathcal{F}_E$, where we expect to see a pure negative contact force along the $\boldsymbol{e}_z^E$ axis. At −6 N, $\boldsymbol{e}_z^E$ is indeed the dominant direction of force. We also notice a positive 1.6 N force along the $\boldsymbol{e}_x^E$, which can be partially explained by pitch error, though it remains at a positive value when the pitch error goes to zero.

To investigate error in the momentum-based force estimate, we inspect the converged values for free-flight static hover at different pitch references. In Fig. 6.40,

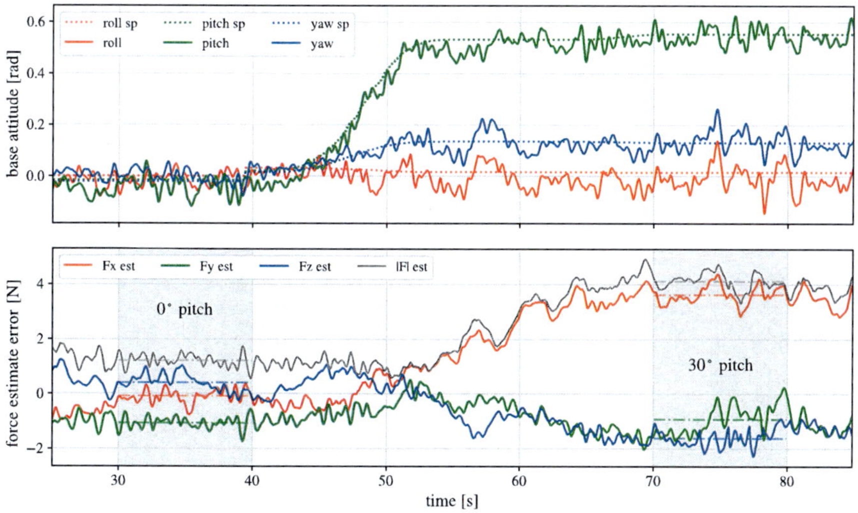

Fig. 6.40 Free flight of the system, with reference pitch changing from 0° to 30°, as shown in the upper attitude tracking plot. Momentum-based wrench estimates are expressed in $\mathcal{F}_E$ below, including the force magnitude in grey. A 10 s shaded region for each pitch is used for statistical analysis, where estimates have converged to their steady-state values

the upper plot shows a flat hover trajectory changing to 30° nose-down pitch. Shaded areas in the lower plot indicate 10 s of steady state force estimates in $\mathcal{F}_E$, which are used for statistical comparison. In the absence of contact, we expect a zero average force, but this is clearly not the case, particularly on the $\boldsymbol{e}_x^E$ axis, which at 30° pitch is closely aligned with gravity. We attribute this phenomenon and the overall rise in force magnitude to actuation model error, including the simplified modelling of double rotor groups and inaccuracies in voltage compensation. These effects will be seen most clearly on the $\boldsymbol{e}_z^W$ axis where the net actuation force aligns against gravity to maintain hover.

We compare the estimated force for the two cases above, along with −30° pitch free-flight hover and rolling-in-contact cases in Fig. 6.41. Force distributions are shown in violin plots, and the corresponding flight orientation for each case is displayed below. The case of −30° pitch gives a reference for steady state force error at the far end of the rolling in contact trajectory where $\boldsymbol{e}_z^E \approx \boldsymbol{e}_z^W$. The force error that we expect to appear along the $\boldsymbol{e}_z^E$ axis also appears substantially in $\boldsymbol{e}_y^E$, which is likely influenced by slight tilt-rotor orientation error. We can conclude from this investigation that the −6 N force estimate while rolling is an underestimate of the true contact force magnitude.

As soon as our desired direction of contact begins to align with gravity, we exhibit compliance to the dominant model error. The generation of consistent omnidirectional contact forces using this axis-selective impedance control framework therefore requires a very good system model. Since this problem results from lumping inter-

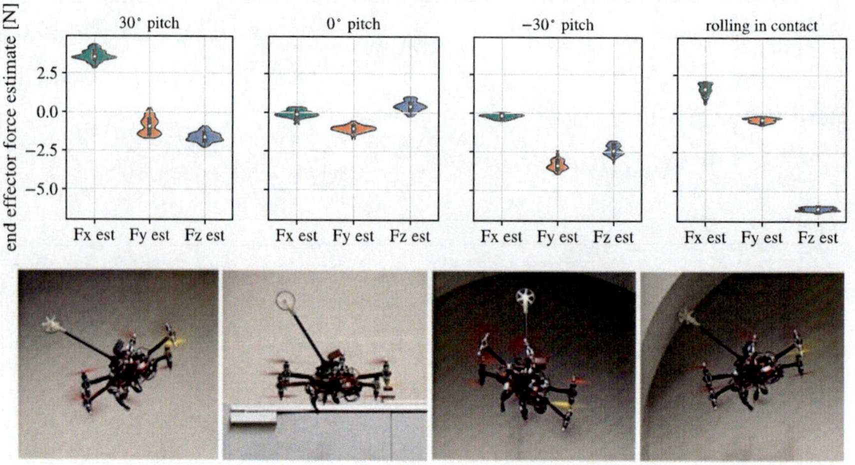

Fig. 6.41 Force estimate data for steady-state free-flight hover is compared for three reference pitches (30°, 0°, −30°), along with the estimate during rolling contact. Images below each plot show the corresponding flight scenario

action forces and model error into a single term, it can also be addressed by using direct interaction force control in our hybrid force-impedance framework.

B: Direct Force Control on an Undulating Surface

In this series of experiments, we evaluate direct force control for point contacts on an undulating surface, using Proto^{V2} with hybrid force-impedance control in the large flight arena. A doubly curved wooden surface is laminated with fibreglass, with dimensions of approximately 1 m × 1.8 m. The surface is mapped using a laser scanner and data is converted for use with a mesh-based planning approach [8].

Figure 6.42 visualizes 42 randomly selected locations on the contact surface. At each contact point, a force trajectory of 10 N perpendicular to the surface is commanded for 5 s. The large number of contact points that are visited during the trajectory allows us to thoroughly investigate performance and repeatability through statistical evaluation of the data.

Measured force data along the $\boldsymbol{e}_z^E$ axis are plotted in Fig. 6.43, where the 5 s time periods of 42 contact commands are superimposed. Oscillations about the 10 N reference are visible for all contact points, though they vary in amplitude. Raw force measurements are processed with a forward-backward 5th order low-pass Butterworth filter, using a 20 Hz cut-off frequency to remove sensor noise while maintaining contact oscillations. A dense collection of data points at 0 N in the first 2 s of contact represents a loss of contact where the end-effector is bouncing against the surface. Due to large initial oscillations and bouncing, we select an evaluation range of 2.3–4.7s, where most trials have gained steady contact and converge towards the

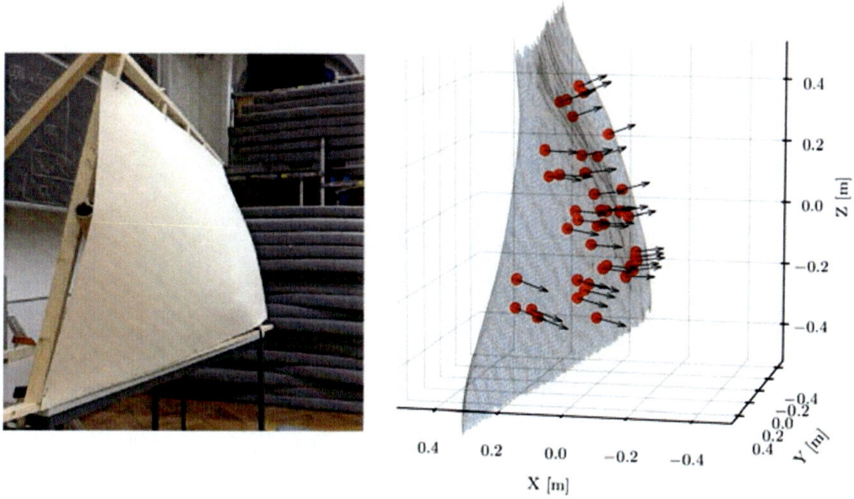

Fig. 6.42 **Left:** Undulating wooden wall used for experiments. **Right**: 42 randomly sampled contact locations plotted as red dots on the mapped mesh surface, where the local surface normal for each point is indicated by a black arrow

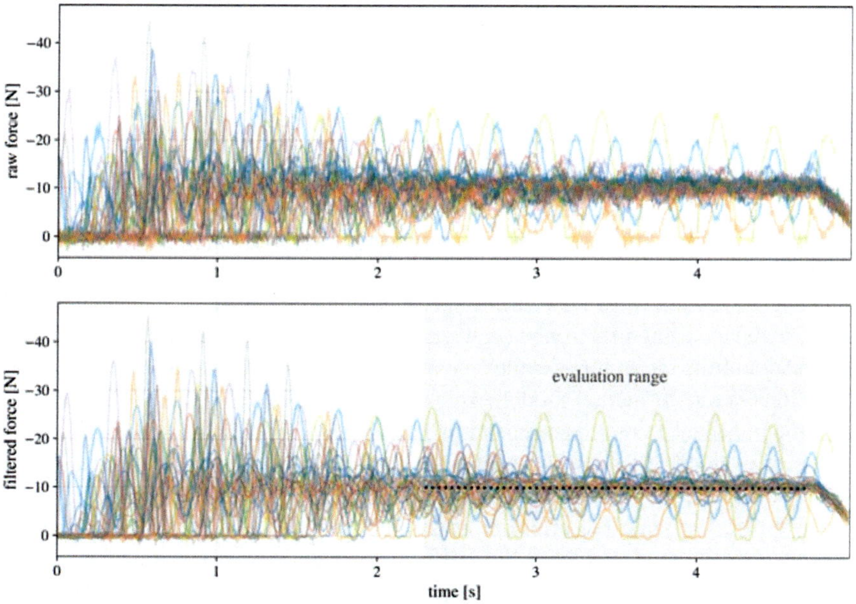

Fig. 6.43 Measured force data along the e_z^E axis of 42 contact trials are plotted for the 5 s duration of a commanded contact force. Raw measurements (upper plot) are processed with a forward-backward 5th order Butterworth filter, using a 20 Hz cut-off frequency for evaluation (lower plot). The shaded region is the range in which force and position error are evaluated for each trial, with the reference force of -10 N shown as a dotted black line

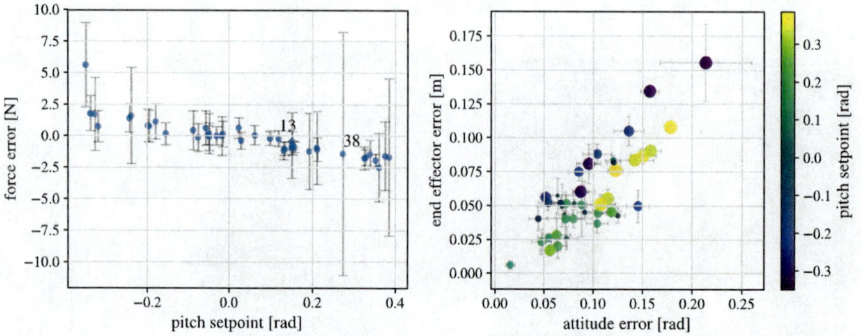

Fig. 6.44 **Left:** Force error along the e_z^E axis is plotted against the pitch reference for each contact, showing a general linear trend. Trial numbers 13 and 38 are labelled. **Right:** The relationship between end-effector error projected onto the surface is plotted against the attitude angle error for each contact, with points coloured and sized by pitch reference value and magnitude

reference force. Force and position error data are evaluated in this range for each contact point.

For evaluation, we consider the following interaction task performance metrics during contact:

- Interaction force error along the e_z^E axis.
- End effector Euclidean error projected onto the local surface plane.
- Total angular error of the attitude (same for end-effector and base).

We are further interested in how the performance is affected by changing pitch reference.

The left plot in Fig. 6.44 shows the relationship between interaction force error and the pitch reference, where mean force error represents an overshoot (negative value) or undershoot (positive value) of the reference force. A linear trend is apparent where a negative pitch reference (upward-tilted end-effector) corresponds with an undershoot of the target force while a positive pitch corresponds with an overshoot. This effect is due to inertial model error, resulting in an initial over- or undershoot from the impedance control command, which is corrected by the force control integral term over time. Low force gains due to noisy force feedback result in slow convergence, and the resulting trend is still visible after 2.7 s of contact.

We further evaluate the effect of magnified end-effector error due to attitude tracking error at the system base. As expected, the right plot of Fig. 6.44 indicates a linear trend where end-effector error projected on the local surface plane grows with increasing total angular error. We observe lower end-effector and attitude error for contacts with zero pitch, and slightly lower end-effector error for positive pitch than for negative pitch references. The latter trend can also be explained by model error in the center of mass location.

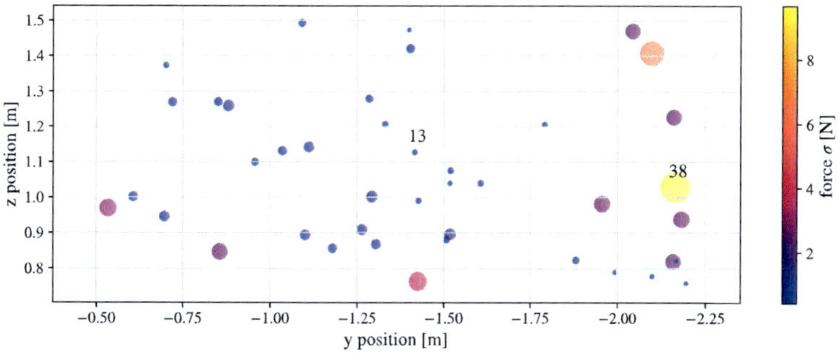

Fig. 6.45 Positions of contact points are projected onto the e_y^W-e_z^W plane. Points are coloured and sized by the standard deviation of e_z^E force for each trial, indicating regions of the surface that exhibit larger oscillations during contact. Trial numbers 13 and 38 are indicated with labels

Standard deviation of the contact force error, σ_f, represents the magnitude of oscillations during contact. Since no consistent trend is visible with respect to the pitch reference, in Fig. 6.45 we plot the contact point locations projected on the e_y^W-e_z^W plane, with colour and size determined by the corresponding σ_f. Agreeing with our observations during experimental flights, contacting certain regions of the structure consistently results in larger oscillations, as seen in the upper right side and left lower edge of the projected surface. Since the surface is made from bent flexible plywood, its local 'springiness' in influenced by the nearest connection to a supporting beam, and varies across the surface. This characteristic means that contact points are not controlled for equivalent interaction properties, but it allows us to investigate the system performance for surfaces with different spring constants.

In Fig. 6.46, we compare commanded and measured forces along the e_z^E axis, along with the corresponding end-effector position for two particular contacts (labelled in Fig. 6.45):

- **Contact** 13: An example of good end-effector position and interaction force tracking.
- **Contact** 38: Point with the largest sustained force oscillations and surface bouncing.

For contact trial 13, we see the measured force converge to the 10 N reference magnitude with small oscillations (± 2 N). The total end-effector deflection into the surface remains in the range of 4 mm during contact, due to relatively high structural stiffness at this point. We note that this distance results from both deflection of the surface and deformation of the foam ball at the end-effector.

For contact trial 38, large force oscillations immediately overshoot the reference force, as expected with a positive pitch reference, interpreted from data in Fig. 6.44. We observe a delay between the commanded force and resulting system motion, resulting in the control signal leading the response by a 90° phase shift. This phenomenon is caused when the system response bandwidth is near a natural frequency

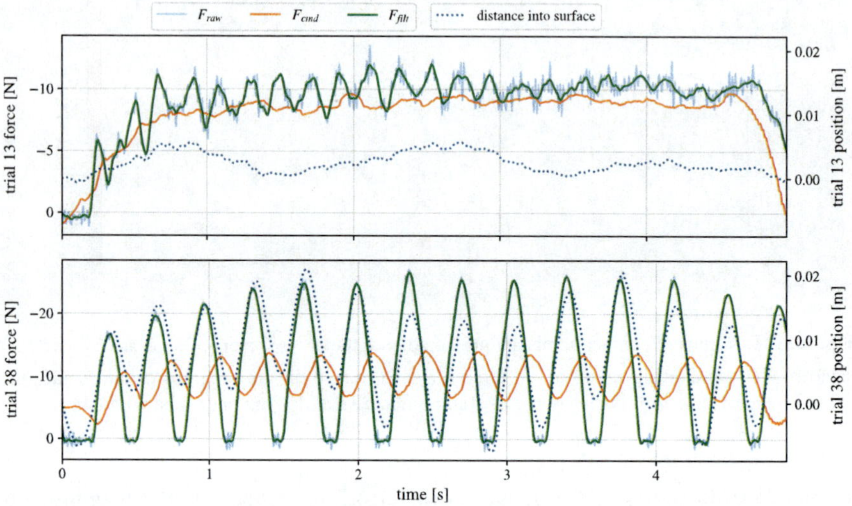

Fig. 6.46 The commanded force is plotted against raw and filtered force measurements along the e_z^E axis for contact points 13 (upper plot) and 38 (lower plot) to show the mechanism of contact oscillations. Position of the end-effector point in the surface normal direction is plotted on a secondary axis

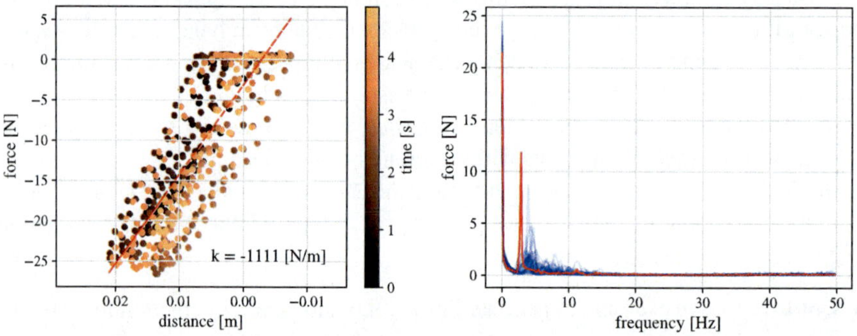

Fig. 6.47 **Left:** Filtered force data along the e_z^E axis is plotted against end-effector position for the entire 5 s duration of contact 38, and a spring constant is estimated by a linear fit of the data. **Right:** Raw force data for the entire duration for each contact is processed with a fast Fourier transform and plotted in the frequency domain, with contact 38 indicated in red which shows a peak at 2.9 Hz

of the combined system, creating sustained oscillations. The main variable that influences the combined system properties between contact trials is the local stiffness of the contact surface.

We investigate the local stiffness of the surface at point 38 in the left plot of Fig. 6.47 by plotting measured force against end-effector deflection along e_z^E for the duration of commanded contact. A linear fit of the data results in a spring constant of 1111 N at this particular location. In the right plot, the raw force data is trans-

Fig. 6.48 Experimental setup for NDT inspection. The end-effector is equipped with a lightweight CSE contact sensor, and contacts a reinforced concrete sample with a known corrosion location. Measurement data is obtained from the on-board sensor and the steel reinforcement by connecting cables

formed into the frequency domain. Data from contact 38 is indicated in red, while the remaining points are plotted in blue. The highest amplitude and lowest frequency non-zero peak is exhibited by contact 38. The remaining trials have smaller force peaks which appear at higher frequencies, corresponding with higher local surface stiffness.

Despite oscillations due to the low-stiffness contact surface, the system showed consistent flight stability over all trials while tracking the nominal desired force during contact. Results indicate that while model improvements should result in better omnidirectional interaction task performance, the system is capable of tracking force trajectories in different orientations. Future work should investigate performance for all possible orientations of the omnidirectional system.

6.3.6 Application: Contact Inspection of Reinforced Concrete

In this experiment, we evaluate the ability of the system to maintain the positional accuracy and force required for measurements with a NDT contact sensor on reinforced concrete structures. We aim to demonstrate the validity of autonomous contact inspection on a sample of reinforced concrete, comparing the inspection results to measurements taken by hand.

The end-effector is equipped with a NDT contact sensor that measures both the electrical potential difference between a saturated CSE and the embedded steel, and the electrical resistance between the sensor on the concrete surface and the steel reinforcement as shown in Fig. 6.48. Electrical potential and resistance results can be used as an indicator for the corrosion state of the steel [9]. A cable is connected to the reinforcement in the concrete structure, and is routed to the flying system via a physical tether to perform the measurements. The concrete specimen used for this experiment has a known corrosion spot at a certain location, to confirm the validity of the results. A constant cover depth over the reinforcement is maintained throughout the block.

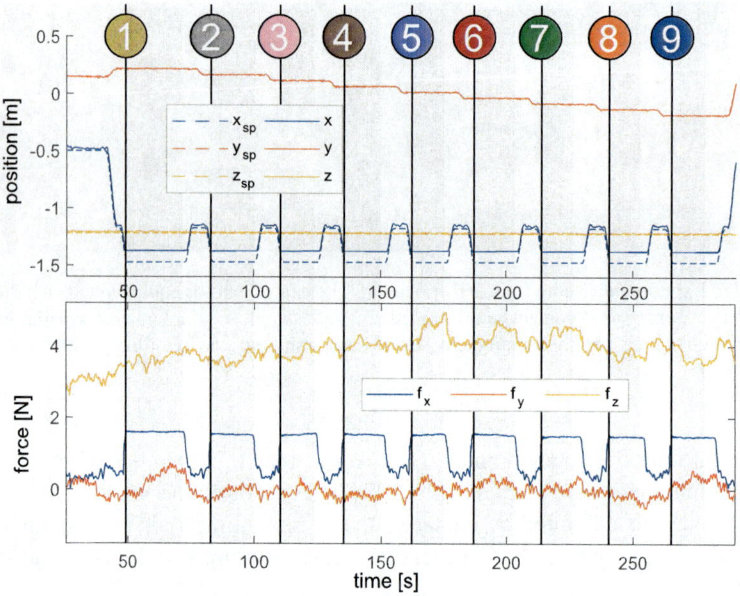

Fig. 6.49 Position tracking and force estimation for tests with a NDT contact sensor on a concrete block, aligned with the x_b-plane. Nine measurements are taken at 5 cm intervals with external state estimation, grey areas indicate contact

We perform two series of experimental trials to validate both **axis-selective impedance control** and **direct interaction force control** in a hybrid force-impedance framework. For both experiments, the concrete block is positioned at a known location, and a trajectory is defined to contact 9 points at 5 cm intervals along the surface. All tests are performed in the large flying area using Vicon fused with onboard IMU data for state estimation.

A: Axis-Selective Impedance Control

Using axis-selective impedance control with Proto^{V1}, we perform the first series of NDT trials. The reference contact point set 10 cm behind the surface of the wall generates sufficient contact force for meaningful measurements. Each point is held for a duration of 15–20 s, and manually commanded to move to the next point when a successful measurement is obtained.

Tracking results in the top plot of Fig. 6.49 show accurate trajectory tracking along all translational axes, except during shaded contact regions, where the $\boldsymbol{e}_z^W$ position is stopped by the concrete block. A low apparent inertia in the $\boldsymbol{e}_z^E$ direction allows for compliant behaviour of the system in contact. The second plot shows forces that arise in the direction of the concrete surface, achieving a value of 1.8 N in the contact phase. A constant offset in force along the $\boldsymbol{e}_z^B$ axis can be seen, which we

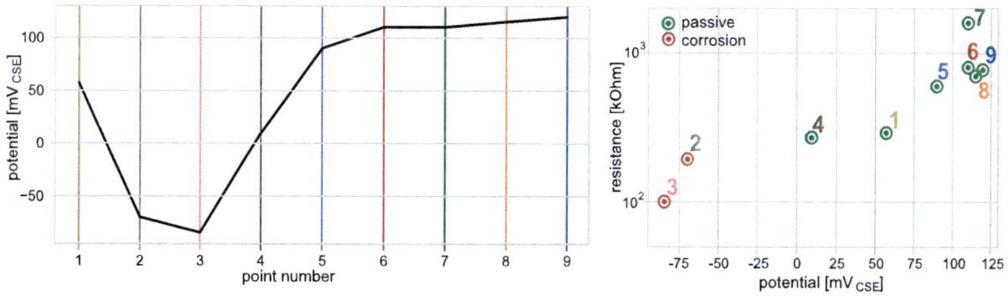

Fig. 6.50 (left) Half-cell potential mapping with 9 points on a concrete specimen reinforced with carbon steel and (right) corrosion state analysis for each point based on resistance and potential measured by an NDT contact sensor

Fig. 6.51 Comparison of potential mapping results for measurements taken during two autonomous test flights, and four sets of manual measurements along a reinforced concrete block sample

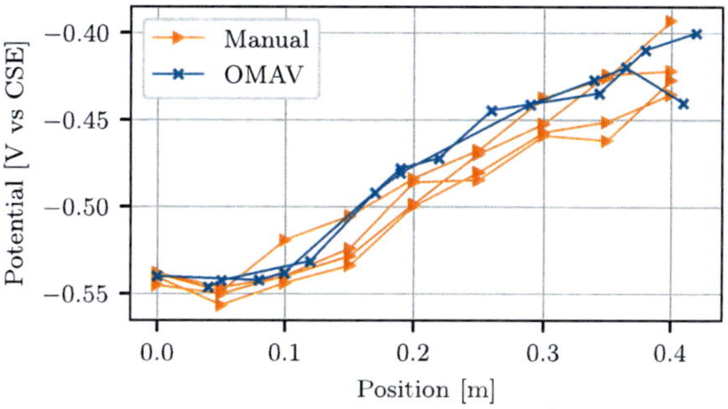

attribute to model error in actuation forces, the system mass and the center of mass offset. Tracking results demonstrate that the controller is robust to this model error in directions with high apparent inertia.

Data collected from the NDT sensor are shown in Fig. 6.50, where corrosion implications are deduced according to [10], and correspond with corrosion at contact points 2 and 3.

A: Direct Interaction Force Control

We perform the second series of NDT trials using Proto^{V2} with hybrid force-impedance control, which uses direct force control during contact phases. Contact reference points are set at prescribed locations on the known surface of the block. Each point is held for a duration of 10 s, during which a reference force of 5 N is included in the trajectory. Compared to the previous trials, the system moves autonomously from one point to the next, without waiting for an operator's command.

Figure 6.51 compares the autonomously measured potentials of two flights with manually measured potentials along the sample, two measured before and two after the flights. Results indicate that the controller can maintain contact between the sensor and the sample for accurate potential measurements.

Table 6.15 Experimental setup for aerial parallel manipulator evaluation

Experiment	Prototype	Controller	Test env.	State est
Section 6.4.1	Manip$^{\Delta}$	inverse kinematic	small arena	Vicon
Section 6.4.2	Proto$^{\Delta}$	PID-inkin	small arena	Vicon-MSF
Section 6.4.3A	Proto$^{\Delta}$	PID-inkin	small arena	Vicon-MSF
Section 6.4.3B	Proto$^{\Delta}$	PID-inkin + FF	small arena	Vicon-MSF

6.4 Macro-Micro Aerial Parallel Manipulator

In this series of experiments, we evaluate the concept of a macro-micro manipulator, in the form of a tilt-rotor omnidirectional flying base combined with a translational delta parallel manipulator. We evaluate the manipulator prototype alone (Manip$^{\Delta}$) and mounted to Proto^{V2} (Proto$^{\Delta}$), to validate the manipulator dynamics presented in Sect. 3.4.3, and confirm the advantages of a macro-micro aerial manipulator for disturbance rejection and end-effector tracking, using controllers presented in Sects. 4.3.1, 4.3.2. We perform the following experiments:

- Section 6.4.1 presents the calibration and validation of the parallel manipulator. A base-to-end-effector calibration is performed, the achievable precision is determined, and control of the servo-motors is evaluated. Complete and simplified dynamic models of the manipulator are validated against ground truth measured reaction forces.
- Section 6.4.2 evaluates the macro-micro system in rejecting disturbances to the flying base compared to a rigid manipulator arm.
- Section 6.4.3 evaluates tracking of a fast end-effector trajectory for the macro-micro system with and without adding a dynamic compensation term, and compares this performance to a rigid manipulator arm. Simulation and real flight results for the dynamic compensation term are evaluated to explore the limitations of this approach.

The experimental setup for this series of test flights is described in Table 6.15. All tests are performed in the small flying arena (see Sect. 6.1.1 for details) equipped with a motion capture system. The Delta arm is rigidly mounted to the base of the omnidirectional flying robot. Reflective marker constellations are installed on the flying base for state estimation, and on the manipulator for ground truth position data and for the initial calibration of the manipulator. State estimation, control, and servo feedback operate at 200 Hz.

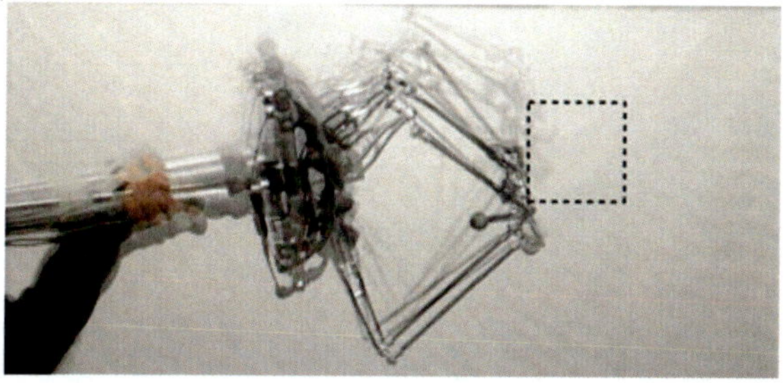

Fig. 6.52 Experimental dynamic model validation by dynamically moving the base while the end-effector tracks a fast trajectory (top image). Plots visually compare measured forces and torques with predicted values from the simplified and complete dynamic models, expressed in $\mathcal{F}_D$ for a short segment of the test

6.4.1 Parallel Manipulator

In this series of experiments, we **validate the reaction wrench** predicted by simplified and complete models of the dynamic parallel structure. Comparing these results, we justify the choice of using a simplified model for the system control. Following validation, we calibrate the manipulator mounted to the base of the flying robot in preparation for flight tests. The end-effector position calculated by forward kinematics is evaluated against ground truth data, since this value is used in the controller.

Dynamic Reaction Wrench

Dynamic model validation of the parallel manipulator alone is performed by comparing reaction force estimates with ground truth force data. The delta manipulator is mounted to a 6-axis Rokubi force-torque sensor, where ground truth position of the sensor base is provided by external motion capture data. We command a fast cube trajectory for the end-effector in $\mathcal{F}_W$, while moving the base dynamically by hand (see Fig. 6.52). This set-up allows us to simulate a floating-base parallel manipulator without the flying system.

The complete and simplified dynamic model results are compared with ground truth data from the force-torque sensor measurements. Figure 6.53 shows a segment of the predicted dynamic wrenches overlaid on the ground truth wrench readings. Violin plots on the right compare the error distributions of the simplified and complete model versus the filtered ground truth data, for which statistical values are presented in Table 6.16. Force and torque tracking results show that both dynamic models closely follow the measured reaction wrench, and the difference between the two is very small. From the violin plots we see that the complete and simplified models have very similar error distributions. Mean error for force and torque predictions are small compared the commanded force magnitudes ($<1\%$).

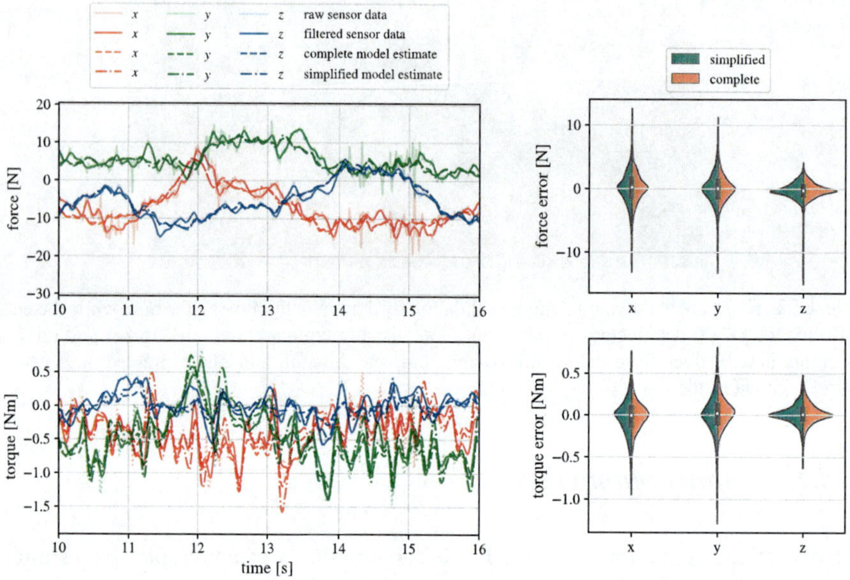

Fig. 6.53 **Left:** For the experimental trial shown in Fig. 6.52, raw and filtered reaction forces measured for a force-torque sensor at the manipulator base are compared with predicted values from the simplified and complete dynamics models of the manipulator. **Right:** Error distributions for both models with respect to filtered sensor data

Table 6.16 Mean, standard deviation, and RMSE of simplified and complete dynamic model error against the force-torque sensor measured ground truth data for 27 s at 200 Hz. All quantities are expressed in $\mathcal{F}_D$

Stat	Model	Force			Torque		
		x	y	z	x	y	z
Mean	**Simplified**	**0.101**	**0.075**	**−0.246**	**−0.026**	**0.006**	**0.016**
	Complete	0.135	0.029	−0.174	−0.014	0.026	0.016
std	**Simplified**	**2.521**	**2.215**	**1.140**	**0.205**	**0.222**	**0.116**
	Complete	2.336	2.039	1.149	0.182	0.179	0.124
RMS	**Simplified**	**2.523**	**2.217**	**1.167**	**0.207**	**0.222**	**0.117**
	Complete	2.340	2.039	1.162	0.182	0.181	0.125

This result confirms that our simplifying assumptions are justified, and the error due to simplification is negligible compared to the remaining model error and noise. Remaining error is likely due to unmodeled dynamics including structural flexibility, vibrations and connected cables, as well as noise and time misalignment from force sensor and joint coordinate measurements.

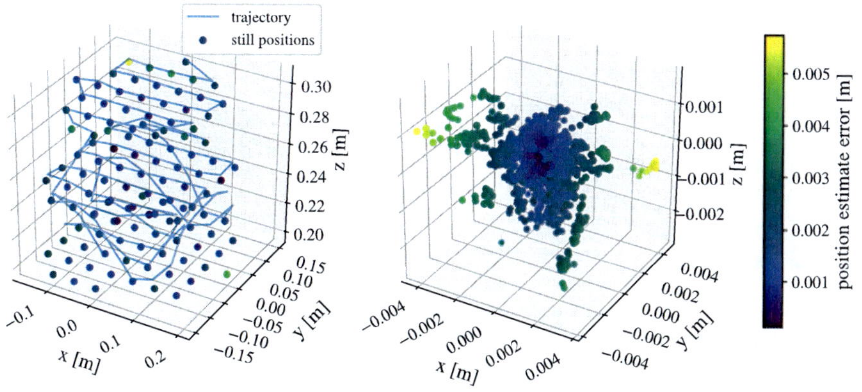

Fig. 6.54 A grid trajectory stops at points covering the end-effector workspace while mounted to the flying base. Ground truth position and measured joint data at stationary points are used to optimize the transform $\boldsymbol{T}_{BD}$

Manipulator Calibration

Manipulator calibration is important prior to flight since the system cannot directly observe the end-effector position. Instead, the controller computed forward kinematics from joint measurements, computed from the delta base frame, $\mathcal{F}_D$.

For kinematic parameters, we use geometric and inertial measurements from a detailed CAD model, after confirming masses and lengths of the produced parts. We then identify the transformation $\boldsymbol{T}_{BD}$ between $\mathcal{F}_B$ and $\mathcal{F}_D$.

A grid trajectory stops at points covering the end-effector workspace while mounted to the flying base. Ground truth positions of $\mathcal{F}_B$ and $\mathcal{F}_D$ are measured by an external motion capture system. The same reflective marker constellations are used later for experimental trials—fused in the odometry estimate for the base, and as ground truth tracking data for the end-effector. Ground truth pose data and measured joint angles at stationary points are used to optimize the transform $\boldsymbol{T}_{BD}$ through an iterative non-linear least-squares problem. Results of the static calibration are presented in Fig. 6.54, where the largest position error norm values are around 5.5 mm, occurring at the outer edges of the workspace.

Applying the transform determined by the calibration above, we then perform an experiment to determine the error in end-effector position estimation we can expect to see when the manipulator is in active motion. The system is commanded to track a static world frame position reference at the end-effector, while the system base is translated and rotated quickly by hand. End effector position error results are shown in Fig. 6.55 in $\mathcal{F}_W$, with error statistics documented in Table 6.17. With added motion of the system, we see a larger error distribution than for the static case. This can be explained by small time misalignment in the ground truth position and joint angle readings which cause more significant error with increasing velocity, and by slight deformations of the structure in dynamic motion. The mean of error distributions

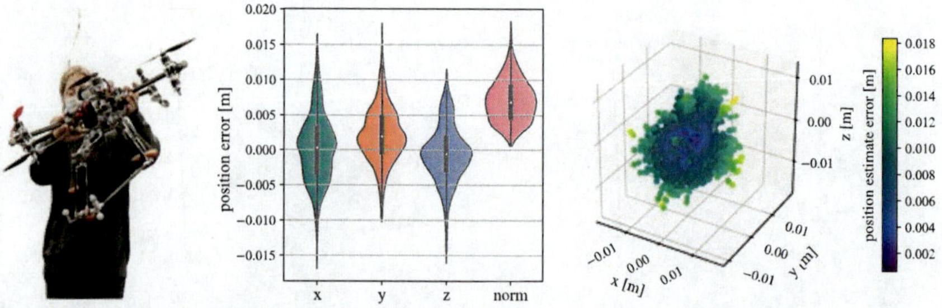

Fig. 6.55 Estimated end-effector position from forward kinematics compared to ground truth. **Left:** The end-effector tracks a constant reference while the base is quickly translated and rotated by hand. **Right:** Violin and 3D error plots for the end-effector position

Table 6.17 Estimation error for end-effector position corresponding with Fig. 6.55

	x	y	z	**Norm**
Mean	0.0002	0.0021	−0.0005	**0.0070**
std	0.0047	0.0039	0.0040	**0.0030**

for each Cartesian axis are small, the largest being 2 mm on $\boldsymbol{e}_y^W$. In $\boldsymbol{e}_x^W$ and $\boldsymbol{e}_z^W$ the mean error is below half a millimeter. With added motion, the mean of the error norm grows to 7 mm, which we can take into account when evaluating ground truth end-effector tracking in flight experiments.

6.4.2 Disturbance Compensation

Due to its low inertia and high speed capabilities, parallel manipulators have been previously used in the state-of-the-art to compensate for disturbances of the flying base, as discussed in Sect. 1.1.3. We design an experiment to evaluate our system for the same type of problem, incorporating the novel omnidirectionality of our tilt-rotor flying base.

A major advantage of the proposed omnidirectional manipulator is its ability to counteract disturbances in 6D, while maintaining any orientation. We show the 6D end-effector tracking capability by commanding a fixed end-effector point with changing orientation, requiring the base to follow a pitching arc trajectory up to 45°. We apply a virtual disturbance force of 10 N to the base along the x-axis of $\mathcal{F}_W$, then compare disturbance rejection performance for the case of a fixed manipulator, and an active delta manipulator (see Fig. 6.56). By comparing to a fixed manipulator, we highlight the advantages of an actuated arm for general disturbance rejection. The tracking performance of the end-effector position, base position and attitude (equal to that of the end-effector), shown in Fig. 6.57, highlight the significant reduction in end-effector position error when using an active delta manipulator. End-effector

Fig. 6.56 Experiments with the base following an arc trajectory up to 45°, and the end-effector tracking a fixed point, while a disturbance is applied for an active (left) and fixed (right) manipulator. Red arrows represent where the virtual disturbance force is applied, while blue circles indicate the desired end-effector position

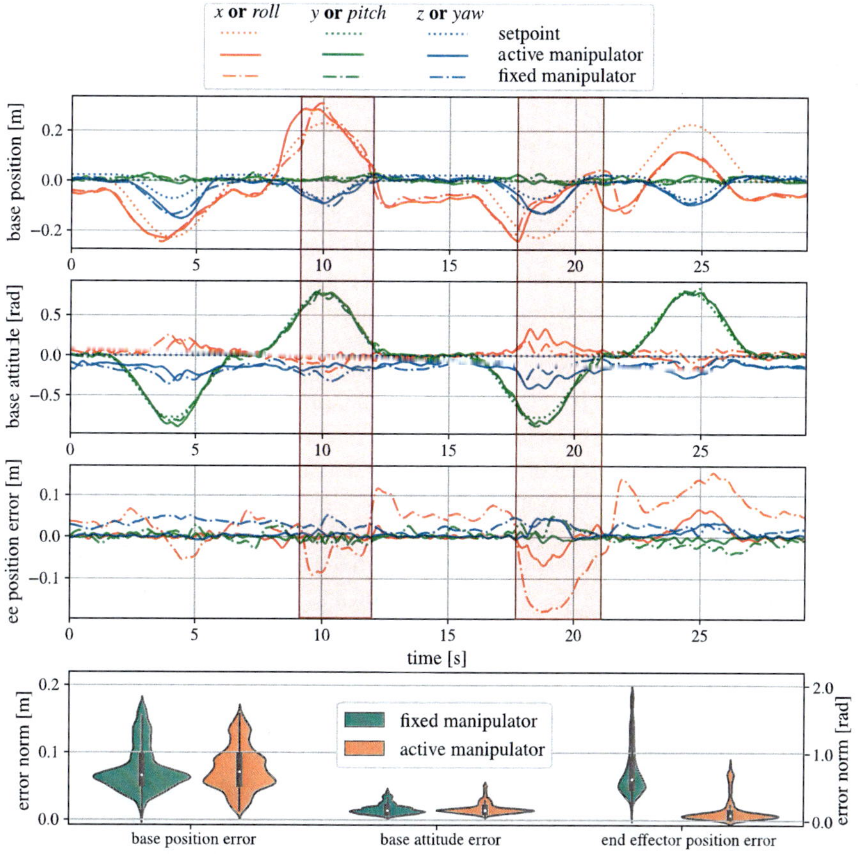

Fig. 6.57 Base position and orientation are plotted relative to their desired trajectory in $\mathcal{F}_W$ in the upper two plots, while the third plot shows end-effector tracking error. Red blocks indicate when the disturbance force is applied. The violin plot (at the bottom) shows base and end-effector tracking error distribution for the fixed and active manipulator cases

error occurring at time 18 and 23 s for the active manipulator case are due to the target point exiting the end-effector workspace.

6.4.3 *Fast End Effector Trajectory Tracking*

In this section, we move beyond state-of-the-art disturbance compensation with parallel manipulators to demonstrate dynamic end-effector tracking abilities from a flying base. A first series of experiments validates the improved performance of macro-micro manipulation compared to a fixed end-effector for different end-effector trajectory speeds with a decoupled control approach. A second set of experiments then evaluates the performance when dynamic compensation of the manipulator is incorporated in the base controller.

A. Decoupled Controller

Focusing on the effect of using an active manipulator for dynamic end-effector tracking, we analyze the relative performance of a fixed and active manipulator for trajectories of different speeds. For the case of a fixed manipulator (simulating a rigidly mounted end-effector), the joint angles are locked to position the end-effector at its nominal workspace center, and a base flight trajectory is commanded to trace the end-effector reference.

We command trajectories which trace the edges of a 0.1 m cube at 1 and 0.5 s per edge, stopping at each vertex. Trajectories are smooth up to the acceleration level. This series of tests is performed in the absence of dynamic compensation terms detailed in Sect. 4.3.2, which corresponds to the state of the art case where dynamic coupling effects are considered negligible.

Ground truth end-effector position tracking in 3D is shown in Fig. 6.59, where the active delta manipulator visibly out-performs it fixed counterpart for both slow and fast cube trajectories. Error distributions are shown as violin plots in Fig. 6.58, confirming the effectiveness of employing an active manipulator for position tracking.

While results for slower trajectories are greatly improved with an active manipulator (norm RMSE 9.4 mm vs 33.2 mm for the fixed manipulator), the faster cube trajectory is not even achievable with a fixed arm, and results in the system becoming unstable due to the base's slower actuation dynamics.

The right tracking plots in Fig. 6.59, as well as at the right two violin plots in Fig. 6.58, we notice degraded performance of the faster cube trajectory tracking. The active manipulator tracks the fast cube trajectory with a norm RMSE of 17.8 mm, almost doubling the value of the slower cube. Since the largest error occurs during fast motion between cube vertices, some of this error can be attributed to estimation error in the end-effector position as described in Table 6.17. Another part of the error can be explained by tracking delay due to position-controlled servos of the manipulator.

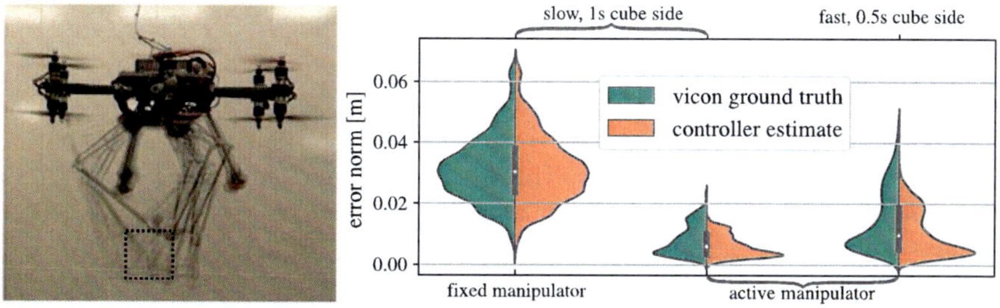

Fig. 6.58 Experimental setup for fast end-effector tracking (left) and Violin plots (right) show the end-effector tracking error for different cube trajectory speeds, for fixed and active manipulators

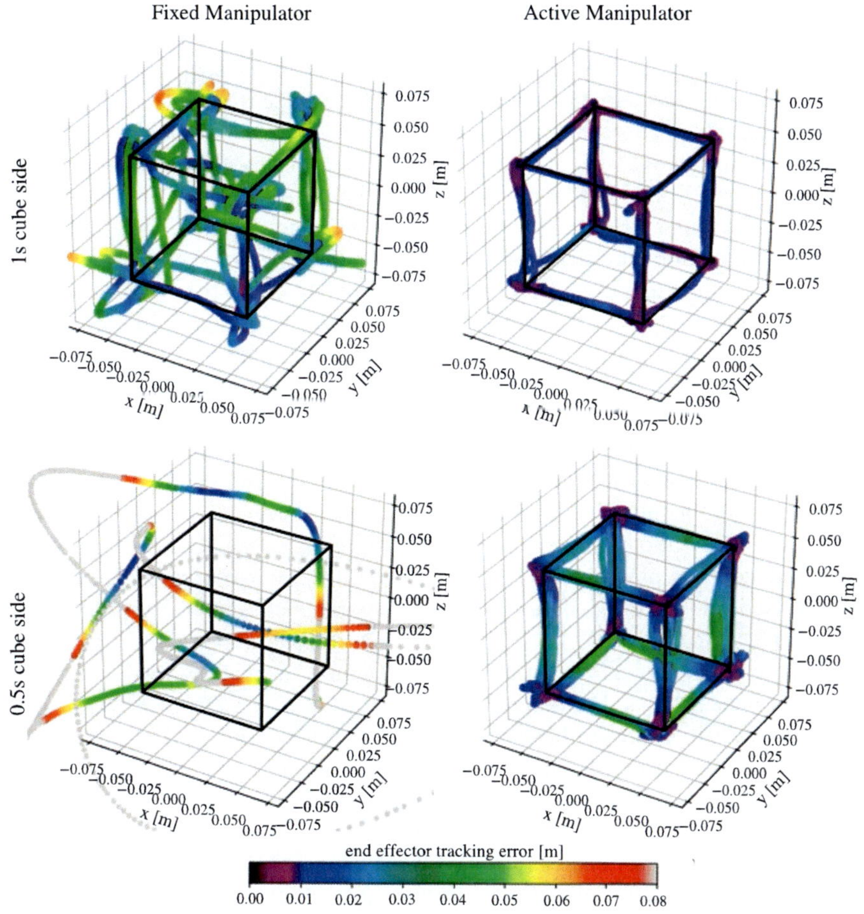

Fig. 6.59 End effector tracking data for 3D points, coloured by the norm of position tracking error. Comparison of fixed arm tracking (left column) with active arm tracking (right column), and of slow 1 s (upper row) and fast 0.5 s (lower row) sided cube trajectory

For the fast cube, we also observe a consistent deviation from the cube edge and overshoot of the vertices, which we not expect from either of the above-mentioned error sources. This final effect can be explained by the non-negligible dynamics of fast end-effector motion, which disturbs the motion of the base and causes a pronounced deviation from the cube edge line. This result motivates using our knowledge of the manipulator dynamics to anticipate and counteract dynamic coupling forces and torques in the base controller.

B. Feed Forward Dynamic Compensation

In this section, we analyze the proposed strategy to compensate dynamic coupling effects during fast end-effector motion. We first experimentally validate the dynamic model derived in Sect. 3.4.3, comparing the predicted dynamic coupling wrench with the readings of a force-torque sensor. We then discuss the tracking performance in both numerical and experimental tests.

In the following two simulation experiments, we command the aerial manipulator to track a world frame oscillating end-effector 'chirp' trajectory of increasing frequency in the range of 0.5–2 Hz while the base reference is fixed, with horizontal attitude. The end-effector trajectory is centered at the nominal delta position in its base frame, with an amplitude of 0.1 m, first horizontally along the x-axis of $\mathcal{F}_W$ and then vertically along the z-axis. Horizontal test results are presented in Fig. 6.60.

The upper plot in Fig. 6.60 overlays the linear acceleration of the excited x-axis in $\mathcal{F}_W$ for the cases with and without dynamic compensation, showing the ability of dynamic compensation to reduce base acceleration towards the desired zero target, particularly for lower frequency arm trajectories. The lower four plots of Fig. 6.60 focus on two sections of the experiment to highlight the interplay of the base controller acceleration terms in $\mathcal{F}_B$ along the oscillating axis for cases without and with dynamic compensation. The middle plots show a region where the arm oscillates at 1 Hz, and the lower plots approaching 2 Hz. The PID term in orange counteracts the position and velocity error caused by the moving arm, but cannot prevent the oscillation. In the middle right plot, the active dynamic compensation effectively reduces oscillations caused by 1 Hz manipulator oscillations. At 2 Hz, however, the lower right plot indicates that the compensation term becomes less effective. We believe that this effect is the result of delays caused in part by numerical differentiation and filtering of the acceleration terms used to compute the compensation term. The effect becomes more prominent when the frequency of the coupling motion exceeds a certain threshold (approx. 2 Hz for this system).

We return to the fast cube trajectory for experimental flights, and compare the performance of the active manipulator without and with the dynamic compensation term. Including dynamic compensation results in more prominent deviation from the reference edges, contrary to our intent, as shown in 3D tracking plots in Fig. 6.61. A number of implementation factors, such as delayed and imprecise base controller action and additional system delays, differentiate simulation results from real experiments.

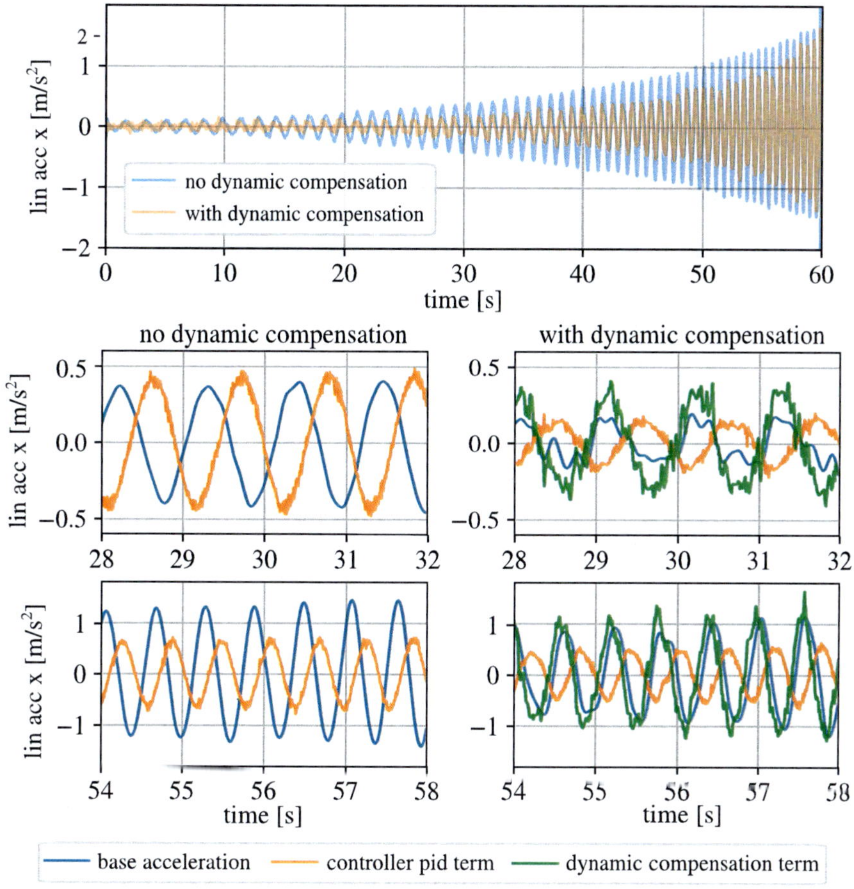

Fig. 6.60 Base acceleration along the excited axis is shown in the upper plot with and without feed-forward dynamic compensation for a horizontally oscillating end-effector trajectory of increasing frequency in simulation. In the lower two plots, a zoomed-in segment of controller PID and dynamic compensation accelerations are plotted over the actual base acceleration along the excited axis

To improve the performance of the system, there are several paths to pursue. If we choose to improve the performance of the current controller, the actuation bandwidth of the system base should be significantly increased, such that the dynamics can be compensated in the short transients we see in the fast cube tracking problem. A more elegant solution for control of the entire system would involve extending the control scheme to include tilting rotor groups in an optimal control scheme, such as SQP, and ultimately to include a receding horizon with MPC. This complex model-based control requires a very good model of the system, and would ideally include torque-controlled motors for tilting rotor groups. Ultimately, both increased control bandwidth and whole-body controllers should be developed to achieve increased end-effector precision and overall task performance during dynamic operations.

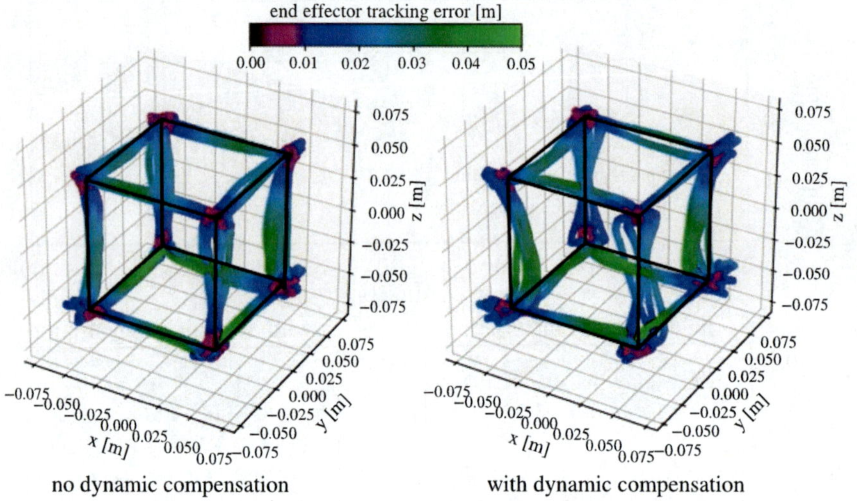

Fig. 6.61 3D end-effector tracking data, coloured by the norm of position tracking error. Tracking of a 1s sided cube trajectory without (left) and with (right) feed-forward dynamic compensation for an active manipulator

References

1. Bodie, K., Taylor, Z., Kamel, M., Siegwart, R.: Towards efficient full pose omnidirectionality with overactuated mavs. In: International Symposium on Experimental Robotics, pp. 85–95. Springer (2018)
2. Allenspach, M., Bodie, K., Brunner, M., Rinsoz, L., Taylor, Z., Kamel, M., Siegwart, R., Nieto, J.: Design and optimal control of a tiltrotor micro-aerial vehicle for efficient omnidirectional flight. Int. J. Robot. Res. **39**(10–11), 1305–1325 (2020)
3. Bodie, K., Brunner, M., Pantic, M., Walser, S., Pfändler, P., Angst, U., Siegwart, R., Nieto, J.: An omnidirectional aerial manipulation platform for contact-based inspection. Robot.: Sci. Syst. **XV**, 15 (2019)
4. Bodie, K., Brunner, M., Pantic, M., Walser, S., Pfändler, P., Angst, U., Siegwart, R., Nieto, J.: Active interaction force control for contact-based inspection with a fully actuated aerial vehicle. IEEE Trans. Robot. (2020)
5. Bodie, K., Tognon, M., Siegwart, R.: Dynamic end effector tracking with an omnidirectional parallel aerial manipulator. IEEE Robot. Autom. Lett. **6**(4), 8165–8172 (2021)
6. Bloesch, M., Burri, M., Omari, S., Hutter, M., Siegwart, R.: Iterated extended kalman filter based visual-inertial odometry using direct photometric feedback. Int. J. Robot. Res. **36**(10), 1053–1072 (2017)
7. Richter, C., Bry, A., Roy, N.: Polynomial trajectory planning for aggressive quadrotor flight in dense indoor environments. In: Robotics Research, pp. 649–666. Springer (2016)
8. Pantic, M., Ott, L., Cadena, C., Siegwart, R., Nieto, J.: Mesh manifold based riemannian motion planning for omnidirectional micro aerial vehicles. IEEE Robot. Autom. Lett. **6**(3), 4790–4797 (2021)
9. Bertolini, L., Elsener, B., Pedeferri, P., Redaelli, E., Polder, R.B.: Corrosion of Steel in Concrete: Prevention, Diagnosis, Repair. Wiley (2013)
10. Jäggi, S., Bürchler, D., Elsner, B., Böhni, H.: Chloride induced corrosion of steel in concrete - Influence of temperature and electrical resistance of concrete. Technical report, Swiss Federal Department of Environment, Transport, Energy and Communications (2005)

Chapter 7
Conclusions

This final chapter summarizes and draws general conclusions on the presented work, and also presents topics of future work that begin where this book leaves off. In a concluding section, we reflect on the problem definition and discuss how the contributions of this research have made steps toward the goal of versatile aerial physical interaction. The following section packages ideas for future work to inspire upcoming research. As a final remark, we comment on the technological readiness of aerial manipulators with an optimistic outlook.

7.1 Conclusions

The content of this book can be summarized with respect to our problem definition and research goals defined in Chap. 2.

After establishing models of tilt-rotor systems and their actuation components, we developed an optimization method to find appropriate morphologies for omnidirectional aerial manipulation based on a design space that is, in reality, a small subspace of possible systems. The idea of versatility had been addressed according to metrics that value highly efficient and omnidirectional flight, though not necessarily at the same time. Our justification of the system's complexity pertaining to these metrics is not to be confused with absolute truth. In many cases, depending on the task, the additional complexity will not be justified. For research purposes, the tilt-rotor system presented in this work has been a very interesting topic of study. As a highly actuated and highly capable system in this context, taking advantage of complexity to demonstrate new control methods can be more insightful than simplifying the system to the needs of a specific task. For some tasks such as simple aerial transport, a quadcopter or fixed-wing system will always be the best solution, though they may not be a good solution for contact inspection of a bridge. Regardless, since autonomous aerial manipulators capable of physical interaction are only

K. Bodie et al., *Omnidirectional Tilt-Rotor Flying Robots for Aerial Physical Interaction*, Springer Tracts in Advanced Robotics 157,
https://doi.org/10.1007/978-3-031-45497-4_7

now emerging for practical use, the suite of tasks for which they will be employed remains unclear. Our job as researchers is still to examine the extent of possibilities, then transfer our knowledge to industry to wrap the ideas into a targeted and efficient solution that benefits society.

Control approaches in this book have divided the control problem into the two modules of an aerodynamic base wrench controller and an allocation method that translates the base wrench to individual actuator commands. This separate control solution has allowed us to implement controllers that are platform-agnostic for omnidirectional flying systems, a more practical result for the aerial robotics research community. From basic PID inverse dynamics tracking control to axis-selective impedance and direct force control for interaction, the development of these controllers for omnidirectional systems is a small step forward using existing concepts in new omnidirectional aerial applications. Integration of force and distance sensing directly in the controller has also allowed us to test the idea of environmental uncertainty, towards our goal that these systems will soon leave the laboratory to find a more purposeful existence in the real world. In many ways, the solution for a base-wrench-controlled system is the simpler form of existing flight controllers for underactuated systems, since the dynamics can be addressed directly without considerations of differential flatness. Instead, some additional control complexity has been explored in actuator allocation, where new forms of singularities must be addressed, and a great degree of coupled over-actuation may be exploited. The instantaneous and differential allocation methods discussed in this book are only a few simple yet practical examples of what can be achieved in this actuation space. Our final contribution in redundant aerial manipulation makes initial steps for flying robots where fixed base and mobile ground robots are already well underway. Our presented method of feed-forward dynamic compensation of an active manipulator presents a minimal solution to a very interesting problem. In general, we seek to realize how the interacting dynamics of a redundant omnidirectional aerial manipulator can be exploited for high-performance aerial physical interaction, beginning with the small step of reaction wrench compensation.

The design work for the complete electro-mechanical tilt-rotor system presented in this book has succeeded in developing a research-grade prototype to test the various control methods. The non-negligible duration of this research has allowed time for design iterations and mechanism development. Despite these iterations, significant improvements can still be made with current technology, particularly if a multidisciplinary team is assembled to dive into implementation details across all software, hardware, and mechanical considerations. In general, the primary immediate system improvement is to upgrade the electronics, to achieve a higher control bandwidth. Recent progress in the size and affordability of high-quality components and sensors have allowed this system to exist in its present form, and we expect that technological development in the coming years will call for a complete re-build, optimizing the system for new and improved components. The design and integration of the parallel manipulator, which began near the end of the thesis work, demonstrates that the additional integration of such a manipulation system is feasible without significant difficulty.

Experimental results have successfully demonstrated the intended behaviours of our system, and knowledge that these can improve with further development of modelling, control and hardware adds to our confidence in the applicability of the proposed methods. Preliminary experimental results for disturbance rejection and fast and precise end-effector tracking confirm the hypothesized effectiveness of the macro-micro solution, in particular of an omnidirectional tilt-rotor base equipped with a delta parallel manipulator. Despite these achievements, several caveats are present which drive the need for future work. Overall system delays, and neglected slower tilt-rotor dynamics limit the effectiveness of controllers that assume instantaneous controllability of an arbitrary base wrench. Admittedly, this control assumption is unrealistic, but presents a baseline for control methods and performance that can be further developed for inclusion of actuation constraints. In the case of macro-micro experiments, attempted compensation of the manipulator dynamics led to reduced end-effector tracking performance, since fast tracking task violates assumptions of the employed base controller and actuator allocation methods. The controllers presented in this book perform at their best after extensive experimental tuning. Our efforts to create a set of tuning parameters that are parametric according to the system mass and inertia are only valid to a certain extent, and some modelling simplifications come back to haunt us, particularly in the case of neglected tilt-rotor dynamics.

7.2 Future Work

The humbling reality about the work presented this book (and perhaps to robotics in general) is that we've only begun to scratch the surface of what can be achieved with omnidirectional aerial manipulation systems. In this section, we touch on a few logical 'next steps' that follow from this work, as hinted at throughout the text.

Improvement of the System Model

A first major body of future work involves removing simplifying assumptions of the omnidirectional tilt-rotor base model, and developing controllers that address a more realistic system representation. The most pressing step is to incorporate correct tilt-joint actuation limits in the actuator allocation, and communicate the resulting wrench limits to the base wrench controller. This requires a detailed study and validation of the rotary joint model for velocity and torque production. The model of propeller transients should also be included, removing the assumption that thrust forces can be achieved instantaneously. These may be generated by fitting a model to empirical data that will differ for acceleration and deceleration of the rotors. The model could vary significantly depending on the implementation of the motor, propeller, and electronic speed controller, and particularly the deceleration model depending on whether or not active braking of the rotors is enabled. Extending beyond model-based control to incorporate reinforcement learning is another option to reduce model error at the component level, while preserving our ability to interpret dynamic capabilities and safe actuation limits of the system. With a good model for tilting motors and rotor thrust transients, we will be better able to examine the wrench control authority of the system, and the effective configuration-based bandwidth of the wrench controller.

Full Body Control

More clever use of the complete system will be enabled with the implementation of full body control. This can be done for the tilt-rotor base with inverse dynamics control of the multi-body system model. With the current prototype, tilt-rotor joints can only be controlled in position or velocity, but ultimately these joints would be controlled in torque. While torque-controlled motors small and light enough to incorporate on a tilt-rotor system of this scale are not yet available, future research may go into the development of these systems. Full body control will also be a critical topic to explore for the floating base parallel manipulator, to incorporate full system dynamics and constraints in a single problem formulation. Through these methods, dynamic interaction tasks with a fast and precise end-effector should see a breakthrough in aerial physical interaction capabilities. Ultimately we would like to use torque-controllable joints for tilt-arm and manipulator joints. This enhancement will enable new dynamic task abilities, and close the gaps between aerial systems and the more advanced development of legged and humanoid robots.

Airflow Interference

As another direction to pursue, one can include a model of rotor group airflow interference, removing the assumption that these effects are negligible. The true aerodynamics of interacting propeller air flows are notoriously complex and difficult to model in real-time, and to model these forces based on traditional methods would involve an exploration in computational fluid dynamics for both static and changing tilt-arm configurations. Machine learning methods may provide a more suitable solution, moving the challenge from highly complex modelling to data collection and processing. Wall and ground effects are another source of unmodelled aerodynamic disturbance, although aerial physical interaction tasks require the system to be close to the environment. In the present work, these effects are not differentiated from contact forces in the momentum-based wrench estimator and lead to inconsistent interaction behaviour when using an impedance control scheme. Estimation and explicit rejection of wind disturbances—as well as prediction of wall and ground effects—could improve overall system performance, e.g. by incorporating additional sensor information.

Interaction Control and Safety

Further improvements to interaction control can be made with the explicit inclusion of contact constraints. These will depend on the contact surface characteristics such as geometry and friction, and offer interesting new avenues of research in perceptive interaction using tactile sensing. Extending beyond the point end-effector model to a plate or gripper will allow the tracking of end-effector torques, and therefore arbitrary wrench tracking at end-effector. The controllers presented in this work do not include safety guarantees during interaction, which will become increasingly important as omnidirectional aerial manipulators become more widespread. The development of passive controllers is important for safe operation in unknown or dynamic environments especially when aerial physical interaction is required. Additional progress in system safety should be made with the fail-safe control development for tilt-rotor systems. The high degree of over-actuation poses an interesting problem in system risk analysis.

Prototype System

While only one of many possible morphologies was realized during this work, many other configurations are possible and should be explored where they offer new system advantages. Further design improvements of the current morphology should also be explored. In the current prototypes, joint limits imposed by tilt-arm cable wind-up trade off flight efficiency to unwind propellers. To completely avoid this problem, the infinitely tilting rotor groups described in the design section here should be developed to a state where they can be safely incorporated into a prototype system. Releasing the system size constraint, and creating a larger system will allow the use of more efficient single-rotor groups. Fewer actuators will also reduce the integration complexity, and allow more straightforward adoption of new communication solutions such as UAV-CAN to command the ESCs. With direct velocity control, higher performance and more accurate model-based control can be expected of this solution, and additional feedback will allow for a safer system through health monitoring and failure prediction. A further step in dynamic control will be the direct control of rotor forces, integrating force-torque sensors directly beneath the rotor groups for closed-loop control and feedback to the system. The present hardware implementation performs the flight control loop at 250 Hz, which is a very low control bandwidth for flying systems. This limits the gains an performance of our system, which should improve by raising the bandwidth ideally to 1 kHz. Throughout this work, rotor speed and tilt angle feedback was not used due to lack of availability or integration capacity, instead using feed-forward actuator control. Integration of this feedback should result in more accurate control wrench generation and improved control performance.

Practical Extensions for Inspection

Applications to NDT inspection of concrete structures will benefit from improved safety and reliability of the system, and overall performance robustness. For outdoor environments for interaction tasks where collision could be catastrophic, drift in position and yaw over time of the VI state estimator prompts the additional integration of an absolute localization source, such as a GNSS, a total station theodolite or feature-based localization. Future extensions of this work should integrate the control approach with mapping, and generate coverage trajectories for complete structural evaluation. These approaches will also improve system safety by detecting and avoiding obstacles. Autonomous data collection with online interpretation of the measurements can lead to more efficient inspection methods, and a better understanding of the state of our infrastructure.

7.2.1 Technological Readiness of Aerial Manipulators

The multitude of research from academic institutions and industry pushing towards autonomous aerial manipulation spurs a whole suite of questions about the technological readiness of these systems, and the impending impact they will have on our society. Are there already effective applications of aerial inspection robots, and where will they start to be employed? What are the new risks associated with this kind of inspection? Will we see a displacement of traditional inspection workers and if so how soon? We attempt to answer some of these questions below.

A major challenge of porting such a system from a laboratory context to the real world is the need for good state estimation close to the environment. While many of the aforementioned systems operate in controlled environments that harness the precise and reliable capabilities of motion-capture state estimation, this approach is not viable in the field. In outdoor environments, GNSS systems are a default choice for most aerial robot operations away from structures or obstacles, often combined with optical flow stabilization. Since inspection necessarily drives the system to the surface of structures that are often large or overhanging, GNSS systems have limited efficacy due to occlusion and multi-path effects [1]. The inclusion of RTK ground stations for GNSS error correction is an option to enhance the global position estimation, but also relies on the calibration of the reference station. Instead, on-board VIO sensors enable state estimation in complex environments, though the typical implementation does not integrate a global reference, and is subject to drift. Solutions to drift can be achieved using SLAM approaches [2], or locally precise position and orientation relative to the environment may be achieved with a 3D TOF camera. Another solution that relies on precise external sensing is the use of a TST, a common tool for precision surveying. Though it provides an excellent global position reference and may be ideal for certain tasks, this system also requires calibration of the ground station, and has limited ability to track fast movements. In light of these state-estimation concerns, aerial manipulators at this point need further research and development before they can be reliably deployed in the wild, though certain applications may already benefit from this technology in a carefully calibrated framework.

Apart from aerial object transport, which has already made the transition to commercial viability, we expect to see several other commercial aerial manipulation applications arise in the next few years. The most immediate applications of aerial physical interaction technology are likely to be in the realm of contact inspection, including force application for contact sensing without modification of the environment. Demand is high for inspection of metal pipelines and concrete bridges, and as new companies raise the reliability of these systems, the reduced costs of inspection are likely to force the transition.

Reduced inspection costs are partly due to a smaller staff of inspection workers, which may lead to the assumption that a loss of jobs will accompany the growth of robotic inspection. In reality, the robotic systems are not intended to replace human beings, but to act as efficient and intelligent tools to perform the same tasks. While the future has yet to reveal the actual numbers in terms of employment, we may optimistically reflect that the lower cost of inspection operations and the great need due to ageing infrastructure means that significantly more inspection operations can take place. With more operations requiring fewer labourers, the number of jobs may not suffer, though some retraining of inspection workers will be necessary. In addition, a whole new category of employment will emerge in the evaluation and interpretation of this inspection data.

References

1. Counselman, C.C., Gourevitch, S.A.: Miniature interferometer terminals for earth surveying: ambiguity and multipath with global positioning system. IEEE Trans. Geosci. Remote Sens. **4**, 244–252 (1981)
2. Durrant-Whyte, H., Bailey, T.: Simultaneous localization and mapping: part i. IEEE Robot. Autom. Mag. **13**(2), 99–110 (2006)

Appendix

A.1 Kinematics

We use kinematics to describe the relative motion that occurs between bodies in multi-body systems including linear and angular position, velocity and acceleration. Without consideration of forces and inertia, this motion describes the system's kinematic model, and will also be useful in deriving the dynamic model.

A.1.1 Rigid Body Transformations

A **rigid body** is a collection of particles that move together such that the distance between any two particles is maintained at all times. Each particle has a position that evolves as a function of time. The net motion of a body, involving translation and rotation, is called a rigid motion.

Positions

A position in 3D Euclidean space is a vector $\boldsymbol{p} \in \mathbb{R}^3$ defined as a point relative to another point, e.g. point C relative to B can be written as $\boldsymbol{p}_{BC}$. The position must be expressed relative to some reference frame, e.g. $\mathcal{F}_A$, which is then written as ${}^A\boldsymbol{p}_{BC}$. For notational simplicity, $\boldsymbol{p}_{BC}$ is used to mean point C with respect to point B expressed in $\mathcal{F}_B$.

Rotations

Using Lie group theory, $\boldsymbol{R}_{WB} \in SO(3)$ denotes a rotation matrix expressing the orientation of $\mathcal{F}_B$ with respect to $\mathcal{F}_W$. The matrix is formed by the collection of orthonormal basis vectors expressing the target frame in the reference frame.

K. Bodie et al., *Omnidirectional Tilt-Rotor Flying Robots for Aerial Physical Interaction*, Springer Tracts in Advanced Robotics 157,
https://doi.org/10.1007/978-3-031-45497-4

$$R_{WB} = \begin{bmatrix} {}^W e_x^B & {}^W e_y^B & {}^W e_z^B \end{bmatrix} \tag{A.1}$$

Rotation matrices have the convenient properties of the $SO(3)$ group, such as $\boldsymbol{R}^\top \boldsymbol{R} = \boldsymbol{R}\boldsymbol{R}^\top = \mathbb{1}$, $\boldsymbol{R}^\top = \boldsymbol{R}^{-1}$ and $\det(\boldsymbol{R}) = +1$. Matrices can be composed such that

$$R_{WC} = R_{WB} R_{BC} \tag{A.2}$$

The rotations defined and used in this document are **passive rotations**, which are commonly used in kinematics to represent the same point or frame in space in a different reference frame. These are passive, meaning that the object itself does not move. Rotations can also be expressed as **active rotations**, in which case the matrix is an operator that rotates the target within the same reference frame, actively changing its real orientation.

For an intuitive understanding of orientation, **Euler angles** will be referred to in experimental evaluation. Euler angles in this document are chosen as the three quantities (ψ, θ, ϕ), for yaw, pitch, and roll. Their disadvantage for use in omnidirectional control is that their representation is ambiguous, having multiple solutions for the same orientation, and certain configurations result in a gimbal lock, or loss of a degree of freedom in the representation. We derive the rotation matrix from Euler angles in ZYX order, such that

$$R_{WB} = R_{Bz}(\psi) R_{By}(\theta) R_{Bx}(\phi) \tag{A.3}$$

$$\begin{aligned} R_{Bz}(\psi) &= \begin{bmatrix} \cos\psi & -\sin\psi & 0 \\ \sin\psi & \cos\psi & 0 \\ 0 & 0 & 1 \end{bmatrix} \\ R_{By}(\theta) &= \begin{bmatrix} \cos\theta & 0 & \sin\theta \\ 0 & 1 & 0 \\ -\sin\theta & 0 & \cos\theta \end{bmatrix} \\ R_{Bx}(\phi) &= \begin{bmatrix} 1 & 0 & 0 \\ 0 & \cos\phi & -\sin\phi \\ 0 & \sin\phi & \cos\phi \end{bmatrix} \end{aligned} \tag{A.4}$$

The unit **quaternion** is another useful representation of rotation that provides a minimal global parametrization, avoiding singularities that are present in Euler angles. The set of quaternions forms a 4 dimensional vector space $\mathbb{H}$ of the form

$$\boldsymbol{q} = q_0 + q_1\boldsymbol{i} + q_2\boldsymbol{j} + q_3\boldsymbol{k} \qquad \boldsymbol{q} \in \mathbb{H}; \quad q_i \in \mathbb{R}, i = 1, \ldots, 3 \tag{A.5}$$

with the shorthand scalar-vector notation of $\boldsymbol{q} = (q_o, \vec{\boldsymbol{q}})$ where $q_o \in \mathbb{R}$ is the scalar component and $\vec{\boldsymbol{q}} \in \mathbb{R}^3$ the vector component. Unit quaternions are normalized for use as rotations such that $||\boldsymbol{q}|| = 1$. Multiplication of quaternions is distributive and associative, but it is not cumulative. Extensive details on quaternion math can be

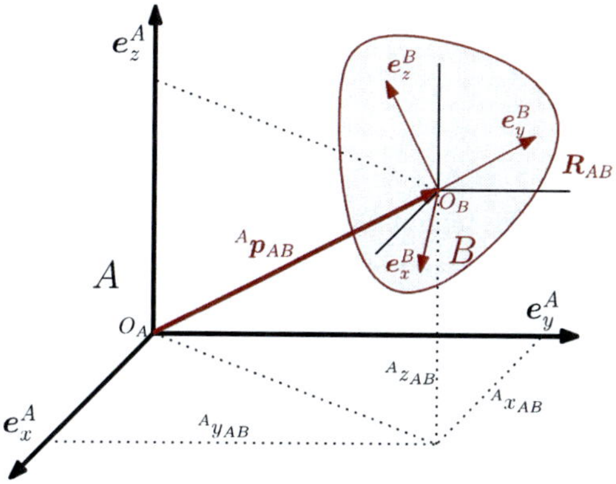

Fig. A.1 Affine transformation of rigid body B with respect to reference frame A: position ${}^A\boldsymbol{p}_{AB}$ of O_B with respect to O_A expressed in $\mathcal{F}_A$, and rotation of $\mathcal{F}_B$ expressed in $\mathcal{F}_A$

found in the original work of Hamilton [1] and excellent summaries are available in various textbooks.

Transformations

We use the general concept of a homogeneous transformation to relate any two reference frames as the combination of a relative position and passive rotation, as shown in Fig. A.1. The transform is minimally expressed as $\{\boldsymbol{R}_{AB}, \boldsymbol{p}_{AB}\} \in SE(3)$, where $SE(3) = \{SO(3) \times \mathbb{R}^3\}$ is the special Euclidean group for 3D space. The homogeneous transformation matrix $\boldsymbol{T}$ is constructed in $\mathbb{R}^{4\times4}$

$$\boldsymbol{T}_{AB} = \begin{bmatrix} \boldsymbol{R}_{AB} & \boldsymbol{p}_{AB} \\ \boldsymbol{0}_{1\times3} & 1 \end{bmatrix} \qquad \boldsymbol{T}_{AB} \begin{bmatrix} {}^B\boldsymbol{p}_{BC} \\ 1 \end{bmatrix} = \begin{bmatrix} {}^A\boldsymbol{p}_{BC} \\ 1 \end{bmatrix} \tag{A.6}$$

This can be used for example to express a point in a different frame, or multiple transformations can be composed my multiplication, with the result remaining in $SE(3)$.

$$\boldsymbol{T}_{AC} = \boldsymbol{T}_{AB}\boldsymbol{T}_{BC} = \begin{bmatrix} \boldsymbol{R}_{AB}\boldsymbol{R}_{BC} & \boldsymbol{R}_{AB}\,{}^B\boldsymbol{p}_{BC} + {}^A\boldsymbol{p}_{AB} \\ \boldsymbol{0}_{1\times3} & 1 \end{bmatrix} \tag{A.7}$$

The inverse of the transformation matrix is formulated as

$$\boldsymbol{T}_{AB}^{-1} = \begin{bmatrix} \boldsymbol{R}_{AB}^\top & -\boldsymbol{R}_{AB}^\top\,{}^A\boldsymbol{p}_{AB} \\ \boldsymbol{0}_{1\times3} & 1 \end{bmatrix} = \begin{bmatrix} \boldsymbol{R}_{BA} & {}^B\boldsymbol{p}_{BA} \\ \boldsymbol{0}_{1\times3} & 1 \end{bmatrix} = \boldsymbol{T}_{BA} \tag{A.8}$$

Angular Velocity

Our general flying base frame $\mathcal{F}_B$ is translating and rotating with respect to the inertial frame $\mathcal{F}_W$. The rotation of a rigid body over time is described by the angular velocity of $\mathcal{F}_B$ with respect to the inertial frame axes, expressed either in the inertial frame as ${}^W\boldsymbol{\omega}_{WB}$ or rotated into the base frame as ${}^B\boldsymbol{\omega}_{WB}$. In the text, we will use the simplified notation ${}^{W}\omega_B$ and ${}^{B}\omega_B$ respectively. It can be shown that the relationship between the angular velocity and the time varying frame rotation matrix can be expressed as

$$ {}^{W}\hat{\omega}_B = \dot{\boldsymbol{R}}_{WB}\boldsymbol{R}_{WB}^{\top}; \qquad {}^{B}\hat{\omega}_B = \boldsymbol{R}_{WB}^{\top}\dot{\boldsymbol{R}}_{WB}. \tag{A.9} $$

Using this relation, if a vector $\boldsymbol{a}$ is expressed in rotating $\mathcal{F}_B$, then the time derivative is expanded as follows:

$$ \frac{d}{dt}\left({}^{B}\boldsymbol{a}\right) = \frac{d}{dt}\left(\boldsymbol{R}_{BW}\ {}^{W}\boldsymbol{a}\right) = {}^{B}\dot{\boldsymbol{a}} + {}^{B}\omega_{WB}{}^{\wedge}\ {}^{B}\boldsymbol{a} \tag{A.10} $$

Regardless of how a rotation is represented, we must be able to relate its time derivative the angular velocity of the rigid body. The Jacobian relating the quaternion to the body angular velocity expressed in $\mathcal{F}_W$ is formulated as ${}^{W}\omega_B = 2\boldsymbol{H}(\boldsymbol{q}_{WB})\dot{\boldsymbol{q}}_{WB} \in \mathbb{R}^3$ where

$$ \boldsymbol{H}(\boldsymbol{q}) = \left[-\vec{\boldsymbol{q}}\ \hat{\vec{\boldsymbol{q}}} + q_0\mathbb{1}_3\right] \in \mathbb{R}^{3\times 4} = \begin{bmatrix} -q_1 & q_0 & -q_3 & q_2 \\ -q_2 & q_3 & q_0 & -q_1 \\ -q_3 & -q_2 & q_1 & q_0 \end{bmatrix}. \tag{A.11} $$

The operator $\hat{\boldsymbol{x}}$, equivalent to $[\boldsymbol{x}]^{\wedge}$ for larger terms, represents the skew-symmetric matrix of vector x.

The Adjoint Transformation

A very useful matrix for transforming motion represented in one coordinate frame to another is the **adjoint transformation** $\boldsymbol{Ad}$ associated with a given homogeneous transformation $\boldsymbol{T}$. After expanding the derivatives of the homogeneous transform, we arrive at the following transformations for linear and angular velocity between frames A and B, either or both of which may be rotating:

$$ {}^{A}\omega_{AB} = \boldsymbol{R}_{AB}\ {}^{B}\omega_{AB} \tag{A.12} $$

$$ {}^{A}\boldsymbol{v}_{AB} = \boldsymbol{p}_{AB} \times (\boldsymbol{R}_{AB}\ {}^{B}\omega_{AB}) + \boldsymbol{R}_{AB}\ {}^{B}\boldsymbol{v}_{AB} \tag{A.13} $$

Which we can rewrite in matrix form to construct the adjoint matrix:

$$ \begin{bmatrix} {}^{A}\boldsymbol{v}_{AB} \\ {}^{A}\omega_{AB} \end{bmatrix} = \underbrace{\begin{bmatrix} \boldsymbol{R}_{AB} & \hat{\boldsymbol{p}}_{AB}\boldsymbol{R}_{AB} \\ \boldsymbol{0}_{3\times 3} & \boldsymbol{R}_{AB} \end{bmatrix}}_{\boldsymbol{Ad}_{AB}} \begin{bmatrix} {}^{B}\boldsymbol{v}_{AB} \\ {}^{B}\omega_{AB} \end{bmatrix} \tag{A.14} $$

The Wrench

The generalized force applied to a rigid body, consisting of a purely linear force component, $\boldsymbol{f} \in \mathbb{R}^3$, and a purely rotational torque component, $\boldsymbol{\tau} \in \mathbb{R}^3$, is referred to as a **wrench**. We represent the wrench $\boldsymbol{w} \in \mathbb{R}^6$ as a stacked vector of force and torque.

$$\boldsymbol{w} = \begin{bmatrix} \boldsymbol{f} \\ \boldsymbol{\tau} \end{bmatrix} \in \mathbb{R}^6 \tag{A.15}$$

The wrench values are relative to the frame in which they are represented, and while the force magnitude does not change with the frame in which it is expressed, the torque does. We consider the coordinate transformation of an applied wrench to be an equivalent wrench of the original, meaning that the transformed wrench applied to the same body at the new coordinates generates the same work for every rigid body motion. The transformation is derived by equating the instantaneous work of the two wrenches applied at frames A and B, where work is the inner product of the twist and the wrench. In the simplified result, using the adjoint matrix to transform the velocity, we find that the coordinate transformation of a wrench is the transposed adjoint matrix:

$$\boldsymbol{w}_A = \underbrace{\boldsymbol{Ad}^{\top}_{\boldsymbol{T}^{-1}_{AB}}}_{\substack{\text{inverse adjoint}\\ \text{transpose}}} \boldsymbol{w}_B \tag{A.16}$$

$$\boldsymbol{Ad}^{\top}_{\boldsymbol{T}^{-1}_{AB}} = \boldsymbol{Ad}^{\top}_{\boldsymbol{T}_{BA}} = \begin{bmatrix} \boldsymbol{R}^{\top}_{BA} & \boldsymbol{0}_{3\times3} \\ -\boldsymbol{R}^{\top}_{BA}\hat{\boldsymbol{p}}_{BA} & \boldsymbol{R}^{\top}_{BA} \end{bmatrix} = \begin{bmatrix} \boldsymbol{R}_{AB} & \boldsymbol{0}_{3\times3} \\ \hat{\boldsymbol{p}}_{AB}\boldsymbol{R}_{AB} & \boldsymbol{R}_{AB} \end{bmatrix} \tag{A.17}$$

A.1.2 Kinematic Model

Here we treat general systems of n multiple rigid body links attached in an open-chain structure by $n-1$ one DoF joints (either rotary or prismatic). An open chain topology stems from a single root link, which can be either fixed or floating. A body-fixed coordinate frame is attached to each link, and a transformation between any two adjacent links can be expressed as a homogeneous transformation that is a function of the connecting joint variable only. Figure A.2 shows modelling examples of fixed-base and floating-base open-chain rigid multi-body systems.

We choose a set of generalized coordinates to completely and concisely describe the system configuration. We aim to choose independent coordinates, such that the representation of the system is minimal, and equal to the number of DoF. For a fixed-based open chain system that is not subject to additional constraints, these coordinates are typically chosen as the set of 1 DoF joint variables. For a floating base system, we further need to describe the pose of the floating base in an inertial frame. A minimal set of coordinates for this representation would include six variables, corresponding to the DoF of a 3D rigid body. To avoid singularities that would arise in time evolution of the system by representing the base orientation with a minimal set of Euler angles, we instead represent the floating base frame as evolving in the 6D SE(3) Lie group. Then, subject to the restrictions of the SO(3) group, we can choose to express orientation equivalently as a rotation matrix or quaternion which are not inhibited by singularities. When using SO(3) in motion, we use the related twist notation to express velocities.

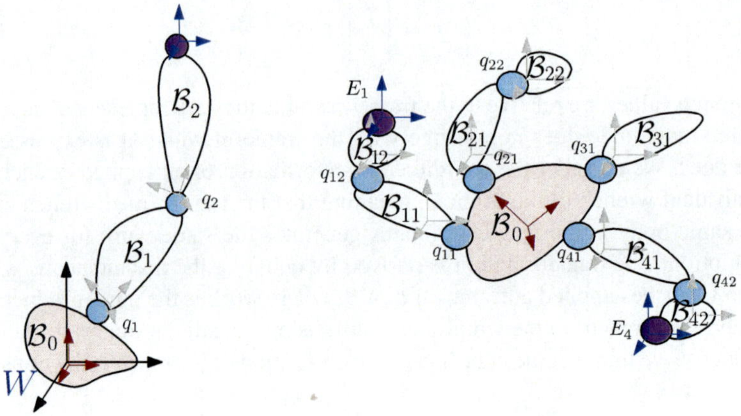

Fig. A.2 Modelling of a fixed base (left) and floating base (right) rigid multi-body system, with rigid body elements $\mathcal{B}$

Usually, we are interested in the transformations between frames where forces are applied (whether by contact, disturbance, or actuation), and the frames where we want to perform motion or force control. Consider a general floating base system with rotary actuators at the end of each open chain of rigid bodies connected by rotary actuators. We express the homogeneous transformation between each rotor group frame $\mathcal{F}_{R_i} \forall i \in \{1 \ \dots \ n_r\}$ and the base frame $\mathcal{F}_B$, in terms of the generalized coordinates $q_{ij} \forall j \in \{1 \ \dots \ m_{ij}\}$ of that chain.

Jacobian

The Jacobian matrix of a vector-valued function of some variable is defined as the matrix of all its first-order partial derivatives. Practically for the motion modelling and control of a multi-body system, we are interested in the Jacobian that maps generalized joint space velocities $\widetilde{\boldsymbol{u}}$ to the velocity of a certain frame, e.g. the end effector linear and angular velocities, or twist.

$$\boldsymbol{\nu}_E = \boldsymbol{J}_E(\boldsymbol{q})\widetilde{\boldsymbol{u}} \tag{A.18}$$

$$\boldsymbol{J}_E(\boldsymbol{q}) = \begin{bmatrix} \frac{\partial \boldsymbol{v}_E}{\partial \widetilde{u}_1} & \dots & \frac{\partial \boldsymbol{v}_E}{\partial \widetilde{u}_{n_q}} \\ \frac{\partial \boldsymbol{\omega}_E}{\partial \widetilde{u}_1} & \dots & \frac{\partial \boldsymbol{\omega}_E}{\partial \widetilde{u}_{n_q}} \end{bmatrix} \tag{A.19}$$

where $\boldsymbol{J}_E(\boldsymbol{q}) \in \mathbb{R}^{6 \times n_q}$ is the end effector Jacobian, for a system with n_q DoFs, and is dependent on the current configuration $\boldsymbol{q}$. The end effector Jacobian is also useful in projecting the wrench $\boldsymbol{w}_E$ that occurs in the end effector frame onto the generalized forces $\widetilde{\boldsymbol{\tau}}$ as

$$\widetilde{\boldsymbol{\tau}} = \boldsymbol{J}_E^{\top} \boldsymbol{w}_E. \tag{A.20}$$

Singularities and Null Space

Considering a system where the independent inputs $\boldsymbol{u} \in \mathbb{R}^{n_i}$ are greater in number than the controllable outputs $\boldsymbol{y} \in \mathbb{R}^{n_o}$, and we have a linear system, we can express their relationship as

$$\boldsymbol{y} = \boldsymbol{B}\boldsymbol{u} + \boldsymbol{c}, \tag{A.21}$$

where $\boldsymbol{B} \in \mathbb{R}^{n_o \times n_i}$ may vary in time based on the robot's configuration. The rank of $\boldsymbol{B}$, representing the maximum number of linearly independent columns, and can be calculated by e.g. singular value decomposition. If the rank is less that n_o, then not all of the outputs are controllable by the inputs, and the system is said to be in a **singular** state. If $\boldsymbol{B}$ is full rank, however, the extra input dimensions indicate the presence of a **null space**. We can make use of the null space projection matrix $\boldsymbol{N} = \mathcal{N}(\boldsymbol{B})$ which satisfies $\boldsymbol{B}\boldsymbol{N} = \boldsymbol{0}$, the simplest projection of which is formulated as

$$\boldsymbol{N} = \mathbb{1} - \boldsymbol{B}^{\dagger}\boldsymbol{B} \tag{A.22}$$

Through this over-actuation, different combinations of inputs can lead to the same output. This input ambiguity should be resolved by a secondary task, whether by minimizing the overall actuation effort in a least squares sense (as is the case with the Moore–Penrose pseudo-inversion $\boldsymbol{B}^{\dagger}$).

A.2 Dynamics

This section provides a brief background for deriving dynamic models of rigid body systems in the rest of this work. We begin by reviewing the Lagrange formulation of dynamical system, then describe the practical reasons for choosing a Newton–Euler approach for flying systems. The content here is largely inspired by [2, 3], where detailed derivations and further content can be found.

A.2.1 Lagrange Formalism

The Lagrange derivation for the equations of motion of a dynamic system uses energy properties for an attached set of particles subject to constraint forces. To begin the Lagrangian formulation, we choose a set of n independent **generalized coordinates** $\boldsymbol{q} = [q_1 \ \dots \ q_n]^{\top} \in \mathbb{R}^n$ which completely describes the configuration of a system with n DoFs. If the system is subject to a set of forces $\boldsymbol{f}_j \forall j \in \{1 \ \dots \ m\}$ applied at points $\boldsymbol{p}_j$, a **generalized force** $\widetilde{\tau}_i$ is described with respect to the $i$þgeneralized coordinate q_i as the projection of all the active forces onto the admissible direction dictated by the variation of q_i.

$$\widetilde{\tau}_i(\boldsymbol{f}, \boldsymbol{q}) = \sum_{j=1}^{m} \boldsymbol{f}_j^\top \frac{\partial \boldsymbol{p}_j}{\partial q_i}, \quad \forall i \in \{1 \dots n\} \tag{A.23}$$

The **Lagrangian** $\mathcal{L}$ is defined as the difference between kinetic $\mathcal{T}(\boldsymbol{q}, \dot{\boldsymbol{q}})$ and potential $\mathcal{V}(\boldsymbol{q})$ energies of the system, and is a function of the configuration $\boldsymbol{q}$ and generalized velocities $\dot{\boldsymbol{q}}$, formulated as

$$\mathcal{L}(\boldsymbol{q}, \dot{\boldsymbol{q}}) = \mathcal{T}(\boldsymbol{q}, \dot{\boldsymbol{q}}) - \mathcal{V}(\boldsymbol{q}). \tag{A.24}$$

The equations of motion are then derived using the **Euler–Lagrange** equation,

$$\frac{d}{dt}\left(\frac{\partial \mathcal{L}}{\partial \dot{\boldsymbol{q}}}\right) - \frac{\partial \mathcal{L}}{\partial \dot{\boldsymbol{q}}} = \widetilde{\boldsymbol{\tau}} \tag{A.25}$$

where $\widetilde{\boldsymbol{\tau}}$ is the vector of generalized forces corresponding to generalized coordinates $\boldsymbol{q}$. If we shift the term $\frac{\partial \mathcal{L}}{\partial \dot{\boldsymbol{q}}}$ to the right-hand side of the equations in Eq. (A.25), the result is a restatement of Newton's law, equating the rate of change of momentum on the left with applied force on the right.

While Lagrange's approach provides a general method for deriving the equations of motion without explicit accounting for internal reaction forces, the configuration space must be expressed in a vector in $\mathbb{R}^n$, where n is the system's total degrees of freedom in configuration space. Since we prefer to parametrize a rigid body on $SE(3)$ instead of $\mathbb{R}^6$, we look instead to the Newton–Euler formulation for a final derivation of the system's equations of motion.

A.2.2 Newton–Euler Formalism

We formulate the Newton–Euler equations of motion starting with Newton's law for a single rigid body, the conservation of linear and angular momentum $\boldsymbol{f} = \frac{d}{dt}(m\boldsymbol{v})$ and $\boldsymbol{\tau} = \frac{d}{dt}(\mathcal{I}\boldsymbol{\omega})$. If we expand these terms, stack linear and angular quantities and perform the derivation in an arbitrarily moving body frame, we generate the Newton–Euler equations of motion in the moving base frame, if the base frame is coincident with the center of mass:

$$\begin{bmatrix} m\mathbb{1}_3 & \boldsymbol{0}_{3\times3} \\ \boldsymbol{0}_{3\times3} & \mathcal{I}_B \end{bmatrix} \begin{bmatrix} \dot{\boldsymbol{v}}_B + {}^B\boldsymbol{g} \\ \dot{\boldsymbol{\omega}}_B \end{bmatrix} + \begin{bmatrix} \boldsymbol{\omega}_B \times m\boldsymbol{v}_B \\ \boldsymbol{\omega}_B \times \mathcal{I}\boldsymbol{\omega}_B \end{bmatrix} = \boldsymbol{w}_B. \tag{A.26}$$

In the case that the CoM, indicated by subscript $(\cdot)_{\circ}$
, is not coincident with the base frame ($||\,{}^B\boldsymbol{p}_{\circ}|| \neq 0$
), we transform these equations from the CoM frame to the $\mathcal{F}_B$.

$$\sum {}^{B}\boldsymbol{w} = \underbrace{\begin{bmatrix} m\mathbb{1}_3 & \mathbf{0} \\ m\,{}^{B}\hat{\boldsymbol{p}}_{\otimes} & \mathcal{I}_B \end{bmatrix}}_{\mathcal{M}_B} \underbrace{\begin{bmatrix} \dot{\boldsymbol{v}}_B \\ \dot{\boldsymbol{\omega}}_B \end{bmatrix}}_{\dot{\nu}_B}$$

$$+ \underbrace{\begin{bmatrix} m\hat{\boldsymbol{\omega}}_B & \mathbf{0} \\ m\,{}^{B}\hat{\boldsymbol{p}}_{\otimes}\hat{\boldsymbol{\omega}}_B & \boldsymbol{\omega}_B \times \mathcal{I}_B \end{bmatrix}}_{\boldsymbol{c}_B} \underbrace{\begin{bmatrix} \boldsymbol{v}_B \\ \boldsymbol{\omega}_B \end{bmatrix}}_{\nu_B} + \underbrace{\begin{bmatrix} m\boldsymbol{R}_{WB}^{\top}\boldsymbol{g} \\ m\,{}^{B}\boldsymbol{p}_{\otimes} \times \boldsymbol{R}_{WB}^{\top}\boldsymbol{g} \end{bmatrix}}_{g_B}. \tag{A.27}$$

The formulation in Eq. (A.27) will become very useful for developing dynamic controllers. We can extend this single-body representation to multiple connected bodies using projected Newton–Euler recursion.

A general rigid-body robot can be expressed as a series of rigid links connected by single-axis joints, either rotary or prismatic. Multi-axis joints can be modelled as a combination of single-axis primitives. Joints are either actuated by a motor (we assume these are controllable in torque, position, or velocity, with feedback in position and velocity), or passive (no applied torque).

The inverse dynamic model of a multi-body system can be written in the same form as a single-body system:

$$\mathcal{M}(\boldsymbol{q})\ddot{\boldsymbol{q}} + \boldsymbol{c}(\boldsymbol{q}, \dot{\boldsymbol{q}}) + \boldsymbol{g}(\boldsymbol{q}) = \tilde{\boldsymbol{\tau}} \tag{A.28}$$

where $\mathcal{M}$ is the generalized inertia matrix, $\boldsymbol{c}$ is a vector containing Coriolis, centrifugal effect, $\boldsymbol{g}$ is a vector containing gravitational effects, and $\boldsymbol{q}$ are the generalized coordinates, with first and second time derivatives $\dot{\boldsymbol{q}}$, $\ddot{\boldsymbol{q}}$. The generalized force vector is $\tilde{\boldsymbol{\tau}}$. We write the direct dynamic model to solve for $\ddot{\boldsymbol{q}}$ as

$$\ddot{\boldsymbol{q}} = \mathcal{M}(\boldsymbol{q})^{-1}(\tilde{\boldsymbol{\tau}} - \boldsymbol{c}(\boldsymbol{q}, \dot{\boldsymbol{q}}) - \boldsymbol{g}(\boldsymbol{q})). \tag{A.29}$$

References

1. Hamilton, W.R.: Elements of Quaternions. Longmans, Green, & Company, London (1866)
2. Hutter, M., Siegwart, R., Stastny, T.: Robot Dynamics Lecture Notes (2017)
3. Siciliano, B., Sciavicco, L., Villani, L., Oriolo, G.: Robotics: Modelling, Planning and Control. Springer Science & Business Media (2010). https://www.springer.com/de/book/9781846286414